Extinct Birds *of* Hawai'i

New Edition

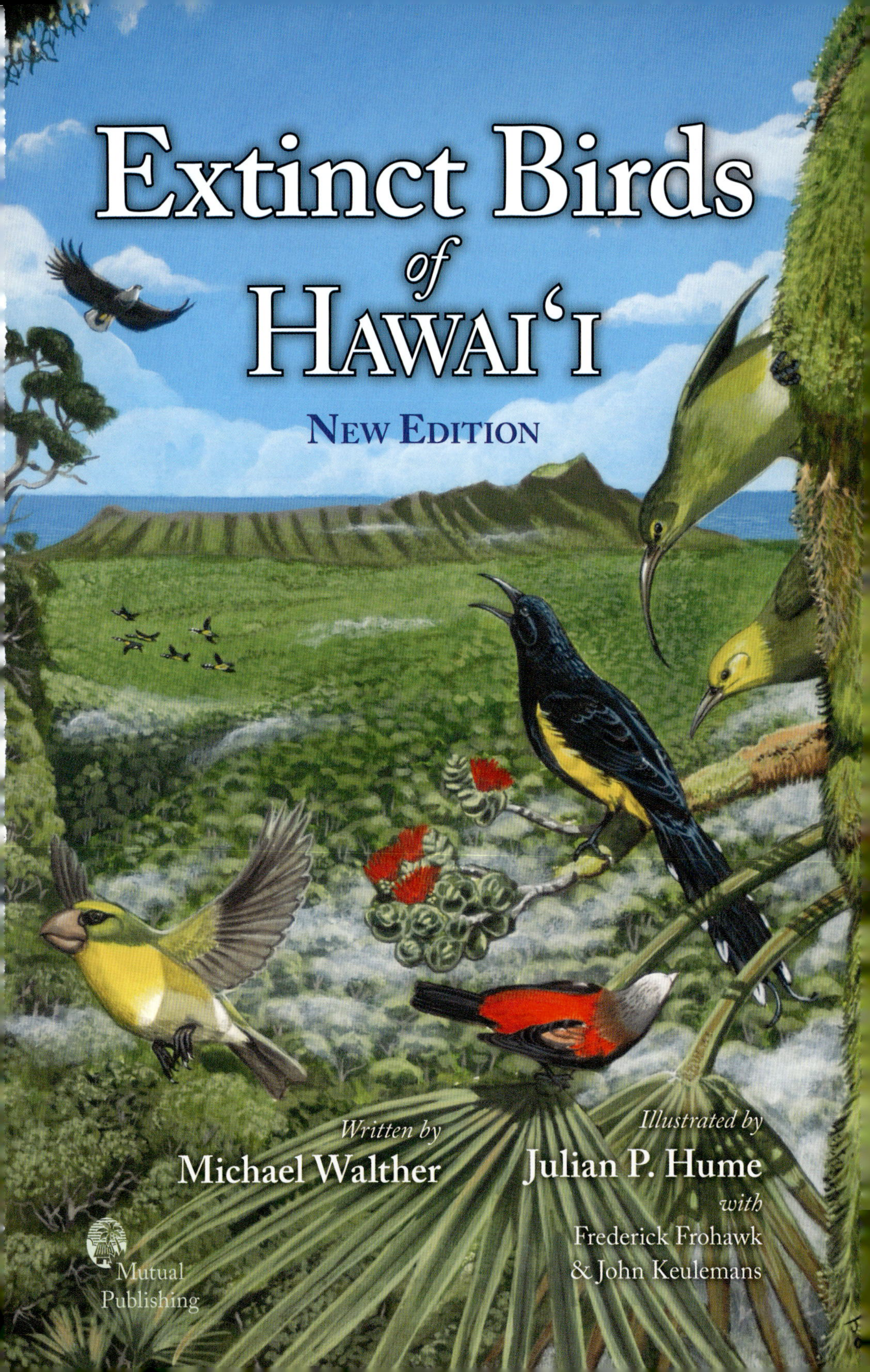

Written by
Michael Walther

Illustrated by
Julian P. Hume
with
Frederick Frohawk
& John Keulemans

Mutual Publishing

Library of Congress Control Number: 2022939859

ISBN: 978-1939487-61-2

First Printing, March 2016
Second Printing, September 2022
Third Printing, September 2024

Mutual Publishing, LLC
1215 Center Street, Suite 210
Honolulu, Hawai'i 96816
Ph: 808-732-1709
Fax: 808-734-4094
info@mutualpublishing.com
www.mutualpublishing.com

Printed in South Korea

A Kioea, *Chaetoptila angustipluma,* drives a Hawai'i 'Ō'ō, *Moho nobilis,* from a flowering 'Ōhi'a, *Metrosideros polymorpha* (Myrtaceae). The Mohoidae were noted for their aggressive behavior toward other birds when flowering trees were in bloom. (JULIAN P. HUME)

Table of Contents

Historical Extinctions

Dedication

Storrs Olson on the Hilo-Volcano Road, Mountain View, Hawai'i. (JULIAN P. HUME)

We would like to dedicate this book to the memory of the late Storrs Lovejoy Olson (April 3, 1944–January 20, 2021), world-leading bird paleontologist, comparative anatomist, historian, and taxonomist, who, along with his first wife, Helen James, made the sensational discovery of fossil birds on the Hawaiian Islands, which more than doubled the number of known species. Over decades, their dedicated research revealed incredible radiations of birds that hitherto were unknown, including giant flightless ducks and geese, flightless ibis, a myriad of flightless crakes, and a new diversity and intra-island distribution of Hawaiian finches that rewrote Hawaiian bird biogeography. Aside from further discoveries of long-legged owls, a harrier, and a diversity of crows—only one species of crow survived into modern times—their most astonishing discovery was of an almost blind, flightless duck! Therefore, what greeted the Polynesians around 1,000 years ago when they first discovered the islands was a bird fauna many times richer than when Captain James Cook arrived in January 1778, the beginning of the modern era.

Storrs and Helen showed that the Hawaiian Islands, already considered a leading hotspot of recent avian extinctions, were only revealing half the story, which illuminates the importance of their palaeontological work. This also has major implications for modern conservation efforts and our ability to protect what remains. The study of Hawaiian birds will always be concomitant with the names of Storrs Olson and Helen James, who follow in the footsteps of the great Hawaiian bird naturalists of the past, like Robert Perkins, George Munro, and Henry Henshaw. For that, we must always be grateful.

—Julian P. Hume & Michael Walther

Preface

The extremely remote tropical islands of the Hawaiian archipelago are the setting of an ongoing bird catastrophe unequalled in world history during the last 700 years. As of 2022, at least seventy-seven bird species and subspecies have become extinct. An additional twenty-five are on the endangered species list, some of which have less than fory-five individuals remaining. There is an urgent need to act to prevent further loss of Hawai'i's biodiversity.

My interest in extinct and endangered birds began more than fifty years ago when I first read about the plight of the California Condor, Peregrine Falcon, Whooping Crane, Ivory-billed Woodpecker, Great Auk, Carolina Parakeet and the Passenger Pigeon. Some of these species represent tragic accounts of extinction, while others are hopeful examples of what can be done if enough concern and money are available to save them.

My purpose and hope in writing this book is to put a spotlight on the devastating extinction crisis in Hawai'i. We urgently need increased public support and funding to halt any further loss of Hawai'i's native bird species. We must learn from the numerous past extinctions before it is too late.

Storrs sieving for fossils (center) in Makauwahi Cave, Kaua'i. (JULIAN P. HUME)

Hawaiian honeycreeper specimens in a collections drawer at the Museum of Zoology.
(© UNIVERSITY OF CAMBRIDGE/MUSEUM OF ZOOLOGY)

Acknowledgments

Thanks to Bennett Hymer, Jane Gillespie, and Courtney Tomasu at Mutual Publishing for the opportunity to write this book and for their excellent assistance with its production.

Extinct Birds of Hawai'i was greatly enhanced by Julian P. Hume's superb paintings. His beautiful portrayals of prehistoric Hawaiian endemic birds, known only from subfossil remains, gives these obscure and forgotten species substance, life, and meaning. I thank him for his kindness, generosity, and interest in this project and for painting many fascinating original works of art specifically for this book.

Special Thanks to Robert L. Pyle and Peter Pyle for creating the excellent monograph, *Birds of the Hawaiian Islands: Occurrence, History, Distribution, and Status*. It is a extremely valuable resource for anyone interested in Hawaiian birds. http://hbs.bishopmuseum.org/birds/rlp-monograph/

Thanks to Frederick Frohawk and John Keulemans for expertly painting many of Hawai'i's historically extinct birds and to Robert Shallenberger for allowing use of one of the few photographs ever taken of the Kaua'i 'Ō'ō. Ms. Kim Evans and Ms. Rene Payne at the Denver Museum of Nature and Science offered kind assistance with images of the Laysan Crake and Laysan Honeycreeper. Thank you to Matthew Cimitile and Marjorie Ziegler for images of Hawaiian extinct bird subfossils and their locations, the Smithsonian for photographs of extinct Hawaiian bird subfossils, and the San Diego Zoo for an image of the last surviving Poʻo-uli.

Thanks to all the scientists, field biologists, and managers working to save Hawaiʻi's forest birds. I especially thank H. Douglas Pratt, Storrs L. Olson, Helen F. James, and Joan Aidem.

To have the time to photograph the birds and plants of Hawaiʻi, to explore the many natural areas of the islands, and to research and write books would not be possible without the many dedicated past and current employees of Oʻahu Nature Tours. Because of their efforts, the company has grown to be a prosperous business; now I can devote many hours to the creative pursuits I find most important. Thanks to them for bringing the message of environmental conservation to thousands of visitors and residents of Hawaiʻi.

I thank my parents, Harold and Marie, for choosing to live near the Everglades in Florida and in a semi-rural suburb of Los Angeles. It was at these special places I developed a love and appreciation for nature. To my brother, Mark, for his generosity and his help in founding Oʻahu Nature Tours and for keeping it going during the difficult early years.

Special thanks to my wife, Cecilia, for her love, friendship, great advice and support, and for building Oʻahu Nature Tours together into a successful company. Her tireless efforts of organizing, managing, and staffing our eco-tour business has enabled me to pursue photographing Hawaiʻi's birds and writing books about nature in Hawaiʻi.

Introduction

The seventy-seven bird species and sub-species extinctions in Hawai'i represent approximately fifteen percent of global bird extinctions during the last 700 years. The catastrophic loss of species that has occurred in this exceedingly small area deserves serious and immediate attention. While the birds of New Zealand and the Mascarene Islands have also suffered great losses, the bird extinction capital of the world is Hawai'i.

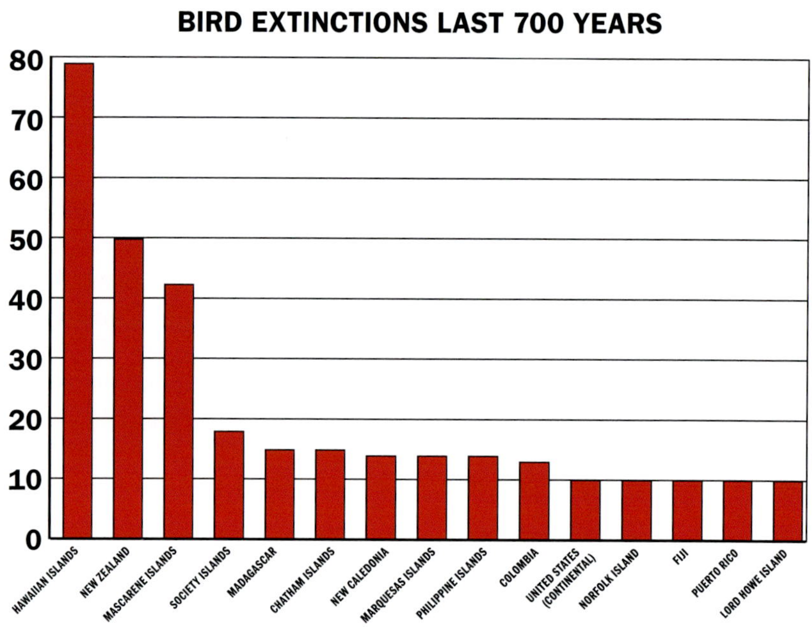

* This graph includes some species and sub-species listed as probably or possibly extinct. Source: *Extinct Birds* by Julian P. Hume.

Several excellent books have been written about global bird extinctions but none have focused only on the extinct birds of Hawai'i. The many causes for species extinction in the Hawaiian Islands include: loss of habitat; reduction of food supply; predation by introduced Roof rats (*Rattus rattus*), mongoose, and feral house cats; competition with non-native birds; destruction of forests by introduced cattle, sheep, goats, and pigs; collection of bird feathers for feather capes, cloaks, helmets, and god images; introduced mosquitoes spreading avian malaria, pox, and other diseases to the native birds; genetic inbreeding; shooting of rare species by museum collectors between 1890-1910; and non-native vegetation displacing native vegetation. For most extinct Hawaiian birds, the exact reason for extinction is unclear and involves a combination of these factors.

Feather cape made with Hawaiian bird feathers (SMITHSONIAN)

The species listed in this book might be revised in the future due to classification of species as sub species and vice versa. There are disagreements about scientific names. For historically extinct Honeycreepers, I have followed H. D. Pratt's 2014 "A Consensus Taxonomy for the Hawaiian Honeycreepers." I have based my tabulations of subfossil species on information provided in *Extinct Birds* (second edition) by Julian P. Hume and from the extensive work done by Storrs L. Olson and Helen F. James. While it is remotely possible a few birds of recently extinct species might have survived, chances for their long-term recovery would be exceptionally slim due to the loss of genetic diversity, the lack of experience with captive propagation for the species, and the continuation of severe limiting forces currently operating in their habitat.

The exact date of extinction is certain for very few birds. The last Laysan Honeycreepers perished during a three-day sandstorm on Laysan Island, April 18-20, 1923. The last known Po'o-uli died in captivity November 26, 2004. Many of Hawai'i's historically extinct bird species were last observed by collectors who tragically shot them for museums. We will never know for certain when the last survivor of many of these birds ended their isolated, lonely existence.

Information on the last observation for most of Hawai'i's bird species is based on the excellent online database *The Birds of the Hawaiian Islands: Occurrence, History, Distribution, and Status* by Robert L. Pyle and Peter Pyle.

All that remains for most historically extinct native Hawaiian forest birds, are old specimens stored in the dark drawers of museums, paintings of the bird often based on the museum specimens, and the written descriptions of scientists, collectors and naturalists who witnessed the living birds. Only a few Hawaiian bird species were photographed before

A flock of cardueline finches arrive on a virgin Hawaiian Island around 4 million years ago. Eurasian rosefinches, *Carpodacus* sp., are the most likely ancestors of the Hawaiian finches. (JULIAN P. HUME)

they became extinct. I have included the superb paintings of Frederick W. Frohawk and Johannes G. Keulemans, both exceptional artists who illustrated two important books written about Hawai'i's birds between 1892 and 1900.

The accounts and descriptions for each species include the exact words that were written as long as 243 years ago. I have chosen not to alter these accounts because I believe they are extremely valuable; they are as important as the specimens themselves. For most of the species a description of plumage, measurements, range, behavior, and songs are included. Due to space limitations I have limited selections of the observations made of the extinct species (please refer to the bibliography). Most of the data for museum collections are from the *History of Endemic Hawaiian Birds: specimens in museum collections* (1979) by Winston E. Banko.

Unfortunately, several species disappeared before any information about their appearance, feeding methods, or voice was recorded. The hope is to utilize the available information to provide an opportunity for the reader to use his or her imagination, and recreate each bird as it once appeared in life.

The native Hawaiian birds that went extinct prior to 1778 were never seen by Western man, and there are no descriptions of a living bird for any of these. Julian's excellent paintings are based on extensive research, but we will never know exactly what these birds looked like. The pre-1778 extinctions have only become known since the first subfossil bones were discovered at Pāhala on the island of Hawaiʻi in 1926. The very important work of Storrs L. Olson, Helen F. James, Joan Aidem, Alexander Wetmore, David Burney, Linda Pigott Burney, Julian P. Hume, and many other dedicated scientists have brought these prehistoric bird extinctions into public awareness, and they have completely changed the old theories of island biogeography and the true extent of the great loss Hawaiʻi's bird species have suffered. For these species, we have even less remaining to help us understand and appreciate what has been lost. They are included in this book because they are a very important part of the history of the extinction of bird species in Hawaiʻi. It is very likely that the actual total will exceed seventy-seven as new subfossil discoveries are made and more species described.

For this second edition, I have included new and updated extinction charts, additional photographs of subfossil bones, museum specimens and feather capes, new and updated paintings by Julian P. Hume, information about the extinct species of Mohoidae that formerly occurred on Maui and details of many historically extinct species last sightings and unconfirmed sightings. Since the first edition was published in 2016, Hawaiian native forest bird species populations continue to decline. Most alarming are the plunging numbers of the ʻAkikiki, ʻAkekeʻe, and ʻIʻiwi on Kauaʻi, and the Kiwikiu and ʻĀkohekohe on Maui. Efforts to eliminate mosquitoes that transmit avian malaria to these rare honeycreepers are increasing. Hopefully Wolbachia-infected mosquitoes will be released into Hawaiian rainforests on Kauaʻi and Maui very soon and this will result in drastically reduced mosquito populations and disease. Please consider donating to the American Bird Conservatory to help fund this extremely important work. https://abcbirds.org/program/hawaii/mosquitoes/

When birdwatchers, nature lovers, and eco-tourists visit Hawaiʻi they are mostly unaware of the tremendous loss of species that has occurred in these remote islands. They do not realize that on some of the islands over 85 percent of the original land bird species are now extinct.

Michael Walther
Honolulu, Hawaiʻi
May 2022

Land Bird Extinctions in the Hawaiian Islands

	HAWAI'I	MAUI	LĀNA'I*	MOLOKA'I	O'AHU	KAUA'I	LAYSAN	NIHOA	STATE TOTALS
TOTAL	25	35	10	26	31	23	4	2	156
EXTANT	12	7	2	4	5	9	1	2	42
EXTINCT HISTORICALY	11	6	7	6	7	5	3	0	45
EXTINCT FROM SUBFOSSIL DATA	2	22	1	16	19	9	0	0	69
TOTAL EXTINCTIONS	13	28	8	22	26	14	3	0	114
% EXTINCT	52%	80%	80%	85%	84%	61%	75%	0%	73%

* Lāna'i fossil species are under-represented due to minimal fossil records at present.

Sources:

Hume, Julian P. and Michael Walters. *Extinct Birds.* Staffordshire: T. & A. D. Poyser, 2012.

Hume, Julian P. *Extinct Birds.* Staffordshire: T. & A. D. Poyser, 2017.

Olson, Storrs L. and Helen F. James. "Descriptions of thirty-two new species of birds from the Hawaiian Islands: Part 1. Non-Passeriformes." *Ornithological Monographs*, 45 (1991): 1-88.

James, Helen F. and Storrs L. Olson. "Descriptions of thirty-two new species of birds from the Hawaiian Islands: Part 2. Passeriformes." *Ornithological Monographs*, 46 (1991): 1-88.

Cape of netted fibre and feathers, collected from Hawai'i in 1778 during Cook's third voyage to the Pacific in the Australian Museum.

Hawai‘i	Status
Pueo (Short-Eared Owl)	Extant
‘Elepaio	Extant
‘Ōma‘o	Extant
Hawai‘i ‘Amakihi	Extant
‘Apapane	Extant
‘I‘iwi	Endangered species
‘Io (Hawaiian Hawk)	Endangered species
‘Alalā (Hawaiian Crow)	Endangered species
Palila	Endangered species
‘Akiapōlā‘au	Endangered species
Hawai‘i Creeper	Endangered species
Hawai‘i ‘Ākepa	Endangered species
Hawaiian Crake (2 ssp.)	Extinct historically
Hawai‘i ‘Ō‘ō	Extinct historically
Kioea	Extinct historically
‘Ō‘ū	Extinct historically
Lesser Koa Finch	Extinct historically
Greater Koa-Finch	Extinct historically
Kona Grosbeak	Extinct historically
Greater ‘Amakihi	Extinct historically
Lesser ‘Akialoa	Extinct historically
‘Ula-‘ai-hawane	Extinct historically
Hawai‘i Mamo	Extinct historically
Giant Hawaiian Goose	Extinct from subfossil data
Giant Scimitar-billed Nukupu‘u	Extinct from subfossil data

KEY
Extant
Endangered species
Extinct historically
Extinct from subfossil data

Maui	Status
Pueo (Short-Eared Owl)	Extant
Hawai‘i ‘Amakihi	Extant
‘Apapane	Extant
‘Alauahio	Extant
‘I‘iwi	Endangered species
Kiwikiu (Maui Parrotbill)	Endangered species
‘Ākohekohe (Crested Honeycreeper)	Endangered species
Maui ‘Ō‘ō	Extinct historically
‘Ō‘ū	Extinct historically
Maui Nukupu‘u	Extinct historically
Maui ‘Ākepa	Extinct historically
Po‘o-uli	Extinct historically
Oloma‘o	Extinct historically
Greater Hawaiian Goose	Extinct from subfossil data
Maui Nui Moa-nalo	Extinct from subfossil data
Stumbling Moa-nalo	Extinct from subfossil data
Maui Flightless Ibis	Extinct from subfossil data
Hawaiian Eagle	Extinct from subfossil data
Kepler’s Crake	Extinct from subfossil data
Severn’s Crake	Extinct from subfossil data
Maui Stilt Owl	Extinct from subfossil data
‘Alalā *(extinct on Maui)*	Extinct from subfossil data
Kioea sp.	Extinct from subfossil data
Narrow-billed Kioea	Extinct from subfossil data
Black Mamo	Extinct from subfossil data
Maui Nui Finch	Extinct from subfossil data
Kiwi Shovel-billed Finch	Extinct from subfossil data
Pololei Shovel-billed Finch	Extinct from subfossil data
Maui Nui Gaper	Extinct from subfossil data
Scissor-billed Koa Finch	Extinct from subfossil data
O‘ahu Koa Finch	Extinct from subfossil data
Wahi Grosbeak	Extinct from subfossil data
Mauka Grosbeak	Extinct from subfossil data
Ridge-billed Finch	Extinct from subfossil data
Maui Nui ‘Akialoa	Extinct from subfossil data

(continued on the next page)

Lānaʻi

- Pueo (Short-Eared Owl)
- ʻApapane
- Hawaiʻi ʻAmakihi
- Olomaʻo
- ʻŌʻū
- Lānaʻi Hookbill
- Maui Nui ʻAkialoa
- ʻAlauahio
- ʻIʻiwi
- Flightless Ibis

Laysan

- Laysan Finch
- Laysan Crake
- Millerbird
- Laysan Honeycreeper

Nihoa

- Nihoa Finch
- Millerbird

Molokaʻi

- Pueo (Short-Eared Owl)
- Hawaiʻi ʻAmakihi
- ʻApapane
- ʻIʻiwi
- Olomaʻo
- Molokaʻi ʻŌʻō
- ʻŌʻū
- Kākāwahie
- Black Mamo
- ʻĀkohekohe (Crested Honeycreeper)
- Hawaiian Eagle
- Hawaiian Harrier
- ʻIo (Hawaiian Hawk)
- Maui Nui Moa-nalo
- Molokaʻi Flightless Ibis
- Molokaʻi Crake
- Molokaʻi Stilt Owl
- Robust Crow
- Laysan Finch *(extinct on Molokaʻi)*
- Nihoa Finch *(extinct on Molokaʻi)*
- Maui Nui Finch
- Maui Nui Gaper
- Ridge-billed Finch
- Kiwikiu (Maui Parrotbill) *(extinct on Molokaʻi)*
- Nukupuʻu
- Maui Nui ʻAkialoa

O'ahu

Pueo (Short-Eared Owl)
O'ahu 'Amakihi
'Apapane
'Elepaio
'I'iwi
'Amaui
O'ahu 'Ō'ō
'Ō'ū
O'ahu 'Akialoa
O'ahu Nukupu'u
O'ahu 'Alauahio
O'ahu 'Ākepa
O'ahu Moa-nalo
Ziegler's Crake
Ralph's Crake
Hawaiian Eagle
Hawaiian Harrier
O'ahu Stilt Owl
Deep-billed Crow
Robust Crow
Kioea sp.
Laysan Finch *(extinct on O'ahu)*
Makawehi Finch
Palila *(extinct on O'ahu)*
Hoopoe-billed 'Akialoa
Straight-billed Gaper
Sickle-billed Gaper
O'ahu Koa Finch
Wahi Grosbeak
King Kong Grosbeak
Ridge-billed Finch

Kaua'i

Pueo (Short-Eared Owl)
'Elepaio
Kaua'i 'Amakihi
'Apapane
'Anianiau
'I'iwi
Puaiohi
'Akikiki
'Akeke'e
Kāma'o
Kaua'i 'Ō'ō
'Ō'ū
Kaua'i 'Akialoa
Kaua'i Nukupu'u
Kaua'i Turtle-jawed Goose
Kaua'i Mole Duck
Kaua'i Stilt Owl
Makawehi Finch
Hoopoe-billed 'Akialoa
Scissor-billed Koa Finch
Pila's Palila
Cone-billed Finch
Kaua'i Palmcreeper

KEY

Extant
Endangered species
Extinct historically
Extinct from subfossil data

World Bird Extinctions in the Last 700 Years

(Sources: *Extinct Birds* by Hume and Walters, 2012 and Hume, 2017)

LOCATION	HISTORICAL	FOSSIL	TOTAL	SPECIES	SUBSPECIES	TOTAL
Hawaiian Islands	35	42	77	72	5	77
New Zealand	21	27	48	36	12	48
Mascarene Islands	37	5	42	42		42
Society Islands	13	5	18	14	4	18
Madagascar	3	13	16	16		16
Chatham Islands	4	12	16	15	1	16
New Caledonia	3	12	15	14	1	15
Marquesas Islands	6	9	15	12	3	15
Philippine Islands	13		13	2	11	13
Norfolk Island	10	2	12	7	5	12
United States (continental)	11		11	7	4	11
Mariana Islands	7	3	10	7	3	10
Lord Howe Island	9		9	5	4	9
Fiji	2	7	9	8	1	9
Australia	9		9	2	7	9
Puerto Rico	3	5	8	7	1	8
Cuba	4	3	7	6	1	7
Colombia	7		7	3	4	7
Solomon Islands	7		7	3	4	7
Canary Islands	2	5	7	6	1	7
Saint Helena		7	7	7		7
Guadalupe	6		6	2	4	6
Jamaica	5	1	6	5	1	6
Bonin Islands	6		6	4	2	6
Cook Islands	2	4	6	5	1	6
Tonga	1	5	6	6		6
Mexico	5		5	3	2	5
Bermuda	2	3	5	5		5
Madeira	2	3	5	4	1	5
Seychelles Islands	4		4	2	2	4
Bahamas	2	2	2	3	1	4

LOCATION	HISTORICAL	FOSSIL	TOTAL	SPECIES	SUBSPECIES	TOTAL
Tasmania	3	1	4	3	1	4
Guadeloupe	4		4	3	1	4
Brazil	3		3	2	1	3
Azores		3	3	3		3
Henderson Island		3	3	3		3
Cayman Islands	2	1	3	2	1	3
Indonesia	3		3	1	2	3
Kiribati	2		2	1	1	2
Haiti	1	1	2	1	1	2
Macquarie Island	2		2		2	2
Morocco	2		2		2	2
Turkey	2		2		2	2
Russia	2		2	1	1	2
Caroline Islands	2		2	2		2
Niue		2	2	2		2
Ascension Island	1	1	2	2		2
Bismarck Archipelago	1	1	2	2		2
Barbados	2		2		2	2
Galapagos	2		2	1	1	2
Peru	2		2		2	2
Martinique WI	2		2	1	1	2
Gambier Islands	2		2	1	1	2
Revillagigedo Islands	2		2		2	2
Congo	1		1		1	1
Tuamotu Islands	1		1		1	1
India	1		1	1		1
Wake Island	1		1	1		1
Assumption Island	1		1		1	1
Samoa	1		1	1		1
Vanuatu	1		1	1		1
Bolivia	1		1		1	1
Auckland Islands	1		1	1		1
Syria, Iraq, Jordan, Saudi Arabia	1		1		1	1
Korea	1		1	1		1
Amsterdam Island	1		1	1		1
Guatemala	1		1	1		1
Austral Islands		1	1	1		1
Tristan Island	1		1	1		1

(continued on the next page)

LOCATION	HISTORICAL	FOSSIL	TOTAL	SPECIES	SUBSPECIES	TOTAL
Ryukyu Islands	1		1	1		1
Amirante Islands	1		1		1	1
Marshall Islands	1		1		1	1
Wallis Islands		1	1	1		1
Kermadec Islands	1		1		1	1
Marie-Galante Island	1		1		1	1
Antigua	1		1		1	1
Costa Rica	1		1		1	1
Juan Fernandez Islands	1		1		1	1
Venezuela	1		1	1		1
Grenada WI	1		1		1	1
Aldabra Island	1		1	1		1
Kenya	1		1		1	1
Saint Lucia	1		1	1		1
Dominican Republic	1		1		1	1
Fernando de Noronha		1	1	1		1
Easter Island		1	1	1		1
Daito Islands	1		1		1	1
St Kitts	1		1		1	1
Hispaniola	1		1	1		1
Channel Islands Ca.	1		1		1	1
San Benito & Isla Cedros Islands	1		1		1	1
Todos Santos Island	1		1		1	1
Tibet	1		1	1		1
Iraq, Iran	1		1		1	1
Burma	1		1		1	1
Virgin Islands		1	1	1		1
Paraguay Argentina	1		1	1		1
Barbuda		1	1	1		1
Christmas Island	1		1	1		1
Campbell Island	1		1	1		1
TOTALS*	**323**	**194**	**517**	**385**	**132**	**517**

* These totals include 88 species and sub-species listed as probably or possibly extinct.

Many of the 54 probably extinct species and subspecies have not been seen in over 100 years. Most of the 34 possibly extinct species and subspecies will, most likely, never be rediscovered and even if they are, the chances for the birds to rebuild into viable populations is very minimal. In Hawaiʻi, five of the seven probably extinct species were officially declared extinct by the United States Fish and Wildlife Service on September 29, 2021.

Hawaiian Bird Subfossils

A Hawaiian Harrier, *Circus dossenus*, pursues a Kākāwahie, *Paroreomyza flammea*, in koa forest on Moloka'i. This hawk had evolved short, broad wings and long legs to hunt small birds in dense forests. (JULIAN P. HUME)

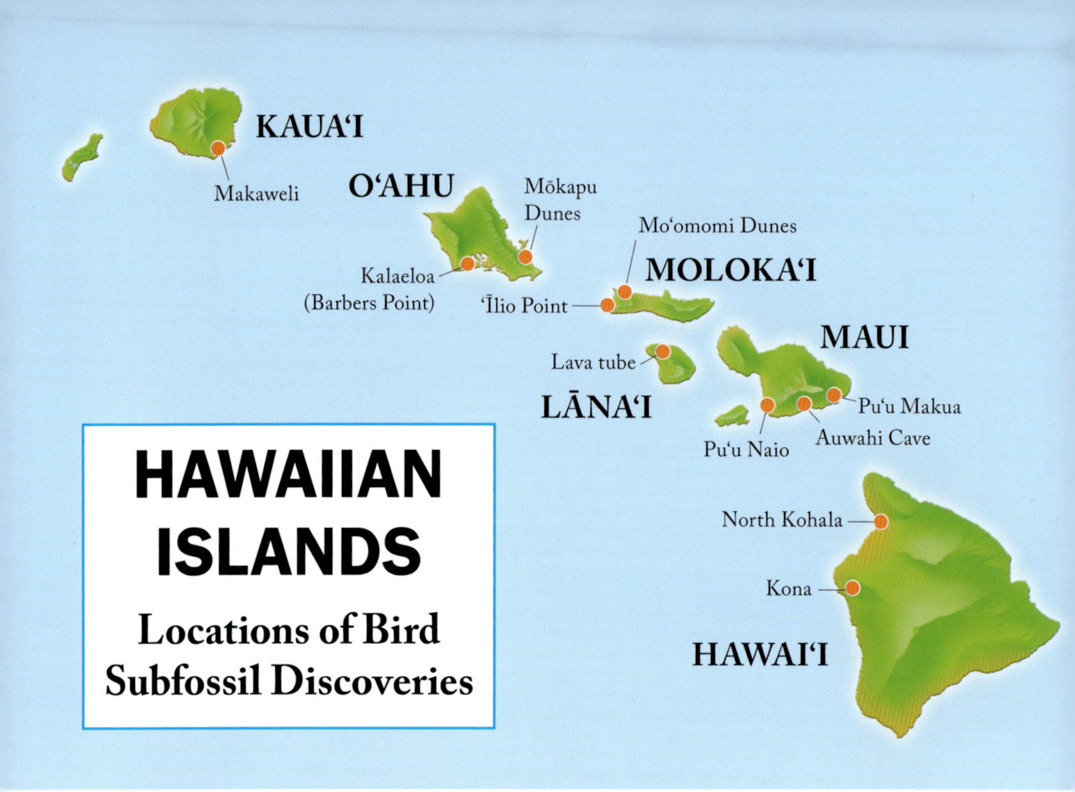

Mo'omomi dunes and beach. (MICHAEL WALTHER)

Bird bones from an extinct giant goose *Branta rhuax* that once inhabited the Big Island of Hawai'i are found in a lava tube. (MATTHEW CIMITILE)

Sinkhole in limestone at Kalaeloa (Barbers Point) O'ahu. (DAVID DUFFY, COURTESY OF CONSERVATION COUNCIL FOR HAWAI'I)

Owl pellet with bones. Mo'omomi Dunes, May 9, 2003. (MICHAEL WALTHER)

Fossil land snail shells at Mo'omomi that are used to help date extinct bird bone deposits. (MICHAEL WALTHER)

Joan Aidem at Mo'omomi Dunes looking for bird subfossils, May 9, 2003. (MICHAEL WALTHER)

Bird bone in sand dunes at Moʻomomi. (MICHAEL WALTHER)

Joan Aidem at Moʻomomi, May 9, 2003. (MICHAEL WALTHER)

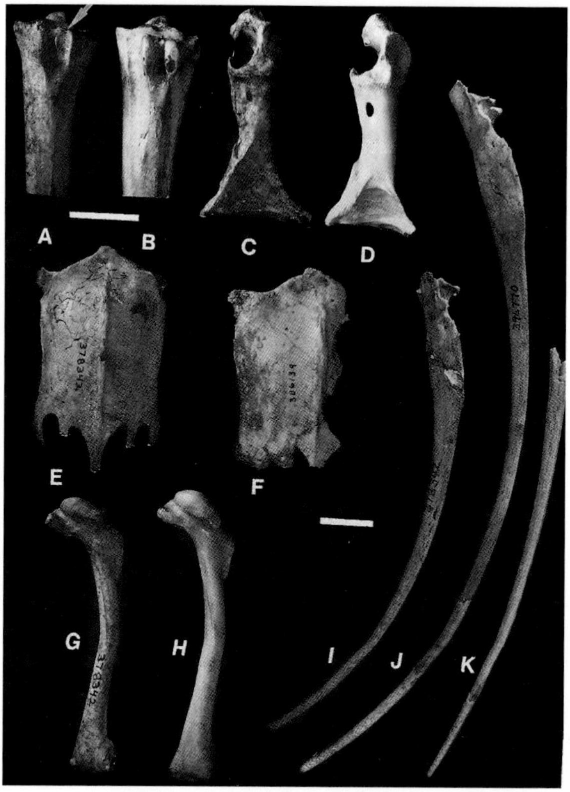

Bones of *Apteribis* spp. from Maui (on left in each pair) compared with those of *A. glenos* from Moloka'i: **A**, **B** proximal ends of right tarsometatarsi in posterior view [**A**: *Apteribis brevis*, new species, holotype (USNM 378342), arrow indicates groove in external calcaneal ridge that is absent in *A. glenos*; **B**: *A. glenos*, holotype (BBM-X 147986)]; **C**, **D**: right coracoids in dorsal view [**C**: *Apteribis brevis*, holotype (USNM 378342); **D**: *A. glenos* (USNM 386119)]; **E**, **F**: sterna in ventral view [**E**: *Apteribis brevis*, holotype (USNM 378342)l **F**: *A. glenos* (USNM 386139)]; **G**, **H**, right humeri in anconal view [**G**: *Apteribis brevis*, holotype (USNM 378342); **H**: *A. glenos* (BBM-X 147240)] **I**, **J**, **K**: mandibles in right lateral view [**I**: *A. brevis,* holotype (USNM 378342); **J**: *Apteribis* sp., Maui (USNM 396770); **K**: *A. glenos* (USNM 426125)]. Scale = 1 cm. (**A** and **B** are enlarged relative to the rest) (COURTESY OF SMITHSONIAN INSTITUTION)

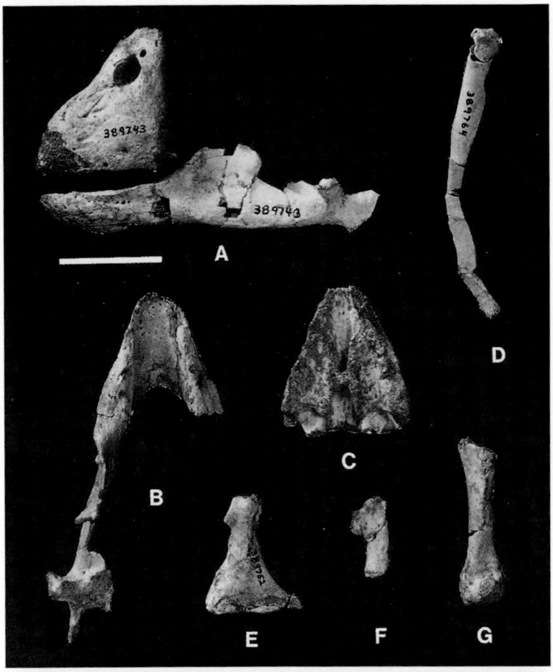

Skeletal elements of *Chelychelynechen quassus,* new genus and species: **A**: rostrum and mandible in left lateral view (USNM 389743, holotype); **B**: mandible in dorsal view (USNM 389743, holotype); **C**: rostrum in ventral view (USNM 389743, holotype); **D**: left scapula in ventral view (USNM 389764); **E**: right coracoid in ventral view (USNM 389762); **F**: left carpometacarpus in external view (USNM 389760); **G**: left humerus (lacking head) in palmar view (USNM 389757). Scale = 3 cm. (COURTESY OF SMITHSONIAN INSTITUTION)

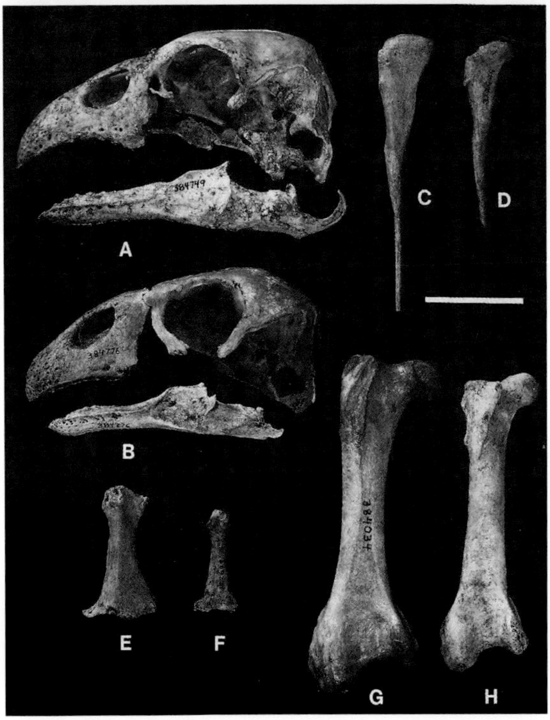

Comparison of bones of the two genera of "toothed" moa-nalos: **A**: skull and mandible of *Thambetochen chauliodous* in left lateral view (one individual with composite number BPBM 159001 and USNM 384749); **B**: skull and mandible of *Ptaiochen pau,* new genus and species, (USNM 384776, holotype) in left lateral view; **C**: left fibula of *T. chauliodous* (BBM-X 146029, holotype) in lateral view; **D**: left fibula of *P. pau* (USNM 322649) in lateral view; **E**: right coracoid of *T. chauliodous* (BBM-X 146029, holotype) in ventral view; **F**: right coracoid of *P. pau* (USNM 322649) in ventral view; **G**: right femur of *T. chauliodous* (USNM 384034) in anterior view; **H**: right femur of *P. pau* (USNM 322649) in anterior view. Scale = 3 cm. (COURTESY OF SMITHSONIAN INSTITUTION)

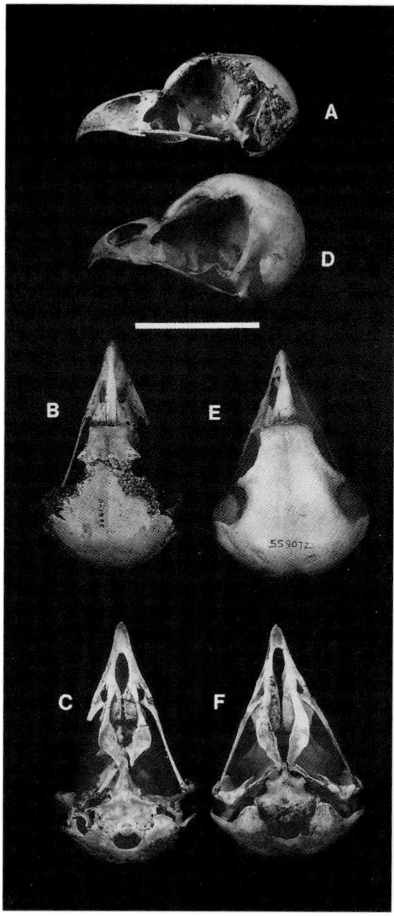

Skull of *Grallistrix geleches*, new genus and species, holotype USNM 386140 (**A**–**C**), compared with *Strix aluco*, USNM 559072 (**D**–**F**): left lateral (**A**, **D**), dorsal (**B**, **E**), and ventral (**C**, **F**) views. Scale = 3 cm. (COURTESY OF SMITHSONIAN INSTITUTION)

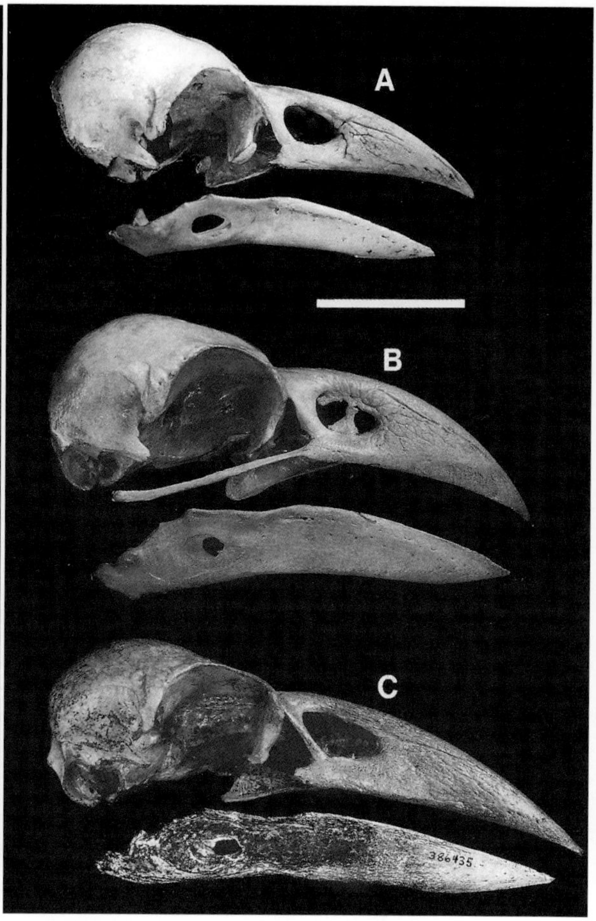

Skull and mandibles of Hawaiian *Corvus* in lateral view. **A**: *C. hawaiiensis* (skull, USNM 177993; mandible, USNM 501638); **B**: *C. impluviatus*, new species, holotype (BBM-X 153652); **C**: *C. viriosus*, new species, holotype (USNM 386435; image of cranium is of the left side printed in reverse). Scale = 3 cm. (COURTESY OF SMITHSONIAN INSTITUTION)

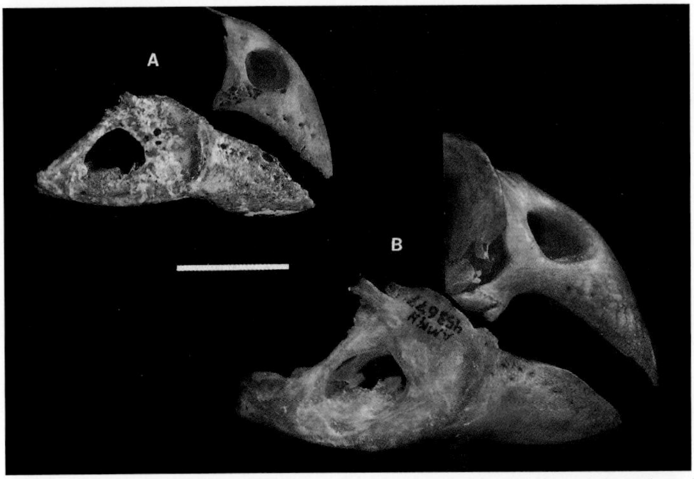

Maxillae and mandibles of *Chloridops wahi,* new species, and *C. kona,* in lateral view, oriented approximately as in life. **A**: *C. wahi* (maxilla = holotype BBM-X 155524, mandible = BBM-X 151278; both images are of the left side printed in reverse); **B**: *C. kona* male (AMNH 453677). Scale = 1 cm. (COURTESY OF SMITHSONIAN INSTITUTION)

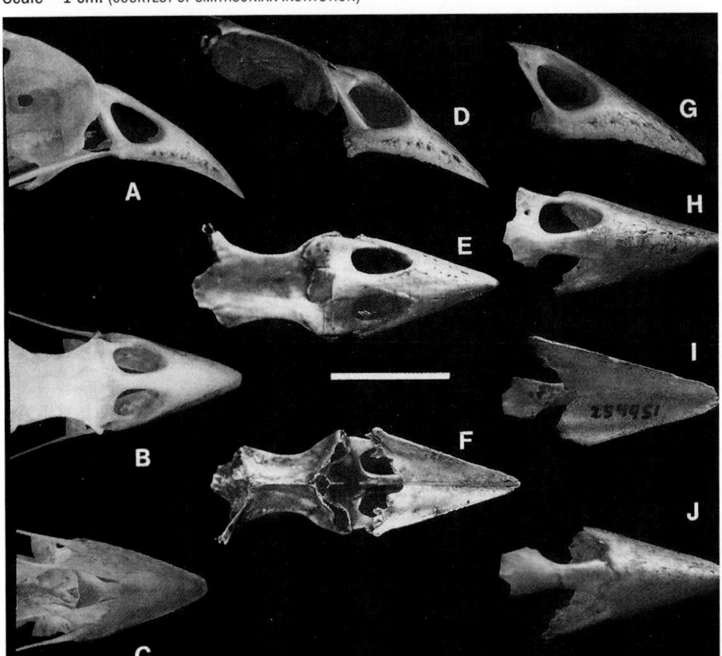

Maxillae of *Melamprosops* and *Xestospiza*, new genus. **A**: lateral, **B**: dorsal and **C**: ventral views of *Melamprosops phaeosoma* (ANMH 810456); **D**: lateral, **E**: dorsal and **F**: ventral views of *X. fastigialis*, new species (USNM 445800; lateral view is of the left side printed in reverse); **G**: lateral, **H**: dorsal and **I**: ventral views of *X. conica*, new species, holotype (USNM 254881; lateral view is of the left side printed in reverse); **J**: dorsal view of *X. conica* (USNM 254951). Scale = 1 cm. (COURTESY OF SMITHSONIAN INSTITUTION)

Extinctions Mostly Before 1778 and Described from Subfossils

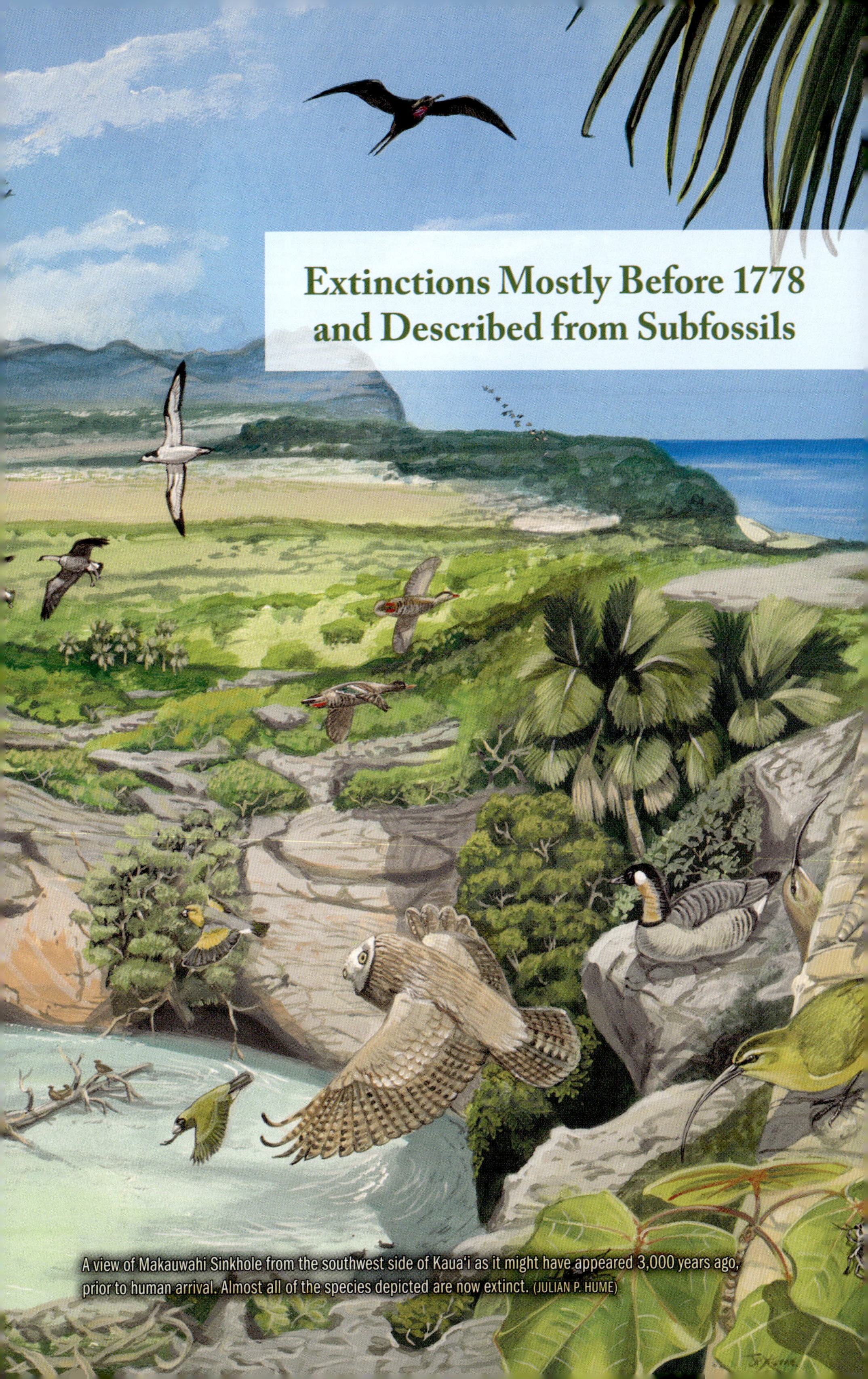

A view of Makauwahi Sinkhole from the southwest side of Kaua'i as it might have appeared 3,000 years ago, prior to human arrival. Almost all of the species depicted are now extinct. (JULIAN P. HUME)

A mixed flock of Greater Hawaiian Goose, *Branta hylobadistes*, (center left) and Nēnē, *Branta sandvicensis*, coexist in dense forest on Haleakalā, Maui. Breeding Laysan Duck, *Anas laysanensis*, (far right) a species thought to only have occurred on remote Laysan Island, and Hawaiian Gallinule, *Gallinula galeata sandvicensis*, (lower right) occur alongside. (JULIAN P. HUME)

Scaled reconstruction of extinct Hawaiian Geese compared with Nēnē, *Branta sandvicensis* (far right). From left: Giant Hawaiian Goose, *Branta rhuax*; Kaua'i Turtle-jawed Goose, *Chelychelynechen quassus*; Maui Nui Moa-nalo, *Thambetochen chauliodous*; O'ahu Moa-nalo, *Thambetochen xanion*; Stumbling Moa-nalo, *Ptaiochen pau*; Greater Hawaiian Goose, *Branta hylobadistes*. (JULIAN P. HUME)

Scaled head studies of extinct Hawaiian Geese compared with Nēnē, *Branta sandvicensis* (center). Clockwise from top left: Giant Hawaiian Goose, *Branta rhuax*; Maui Nui Moa-nalo, *Thambetochen chauliodous*; O'ahu Moa-nalo, *Thambetochen xanion*; Stumbling Moa-nalo, *Ptaiochen pau*; Greater Hawaiian Goose, *Branta hylobadistes*; Kaua'i Turtle-jawed Goose, *Chelychelynechen quassus*. (JULIAN P. HUME)

Giant Hawaiian Goose

Branta rhuax

A very large flightless goose estimated to be on average 1.4 times heavier than the largest subspecies of continental Canada goose (*B. canadensis maxima*) and 4.3 times the weight of the extant Nēnē, *Branta sandvicensis*. The bill was robust and deep and the leg bones were massive but the wings were disproportionately small. Based on mitochondrial DNA analysis, this extinct species evolved from a Canada Goose less than 500,000 years ago.[1]

Geographic range: Island of Hawai‘i. Fragmentary subfossil remains of this species were found in lava tubes located in the districts of Kona and North Kohala.[2]

Subfossil remains: Bishop Museum (Honolulu) and Smithsonian (Washington, D.C.)

Male Giant Hawaiian Geese, *Branta rhuax*, fight over territory in a kūpuka in the foothills of Mauna Loa volcano. A female and juvenile look on while a Western Hawaiian Crake, *Zapornia s. sandwichensis*, runs for cover. (JULIAN P. HUME)

Greater Hawaiian Goose, *Branta hylobadistes,* (left) compared with Nēnē, *Branta sandvicensis*. Note the more robust head and legs and shorter wings in the Greater Hawaiian Goose. This species was in the evolutionary process of becoming flightless. (JULIAN P. HUME)

Greater Hawaiian Goose (Nēnē-Nui)

Branta hylobadistes

This species was related to both the Giant Hawaiian Goose, *Branta rhuax* and the extant Nēnē, *Branta sandvicensis.* It was a large goose with small wings and robust limbs.[3] Uniquely, subfossil remains shows that this species was in the process of losing its flying ability, with some specimens flightless, others volant. The scientific name, *Branta hylobadistes*, means woods-walking goose.

Geographic range: Island of Maui

Subfossil remains: Smithsonian (Washington, D.C.)

Maui Nui Moa-nalo

Thambetochen chauliodous

The flightless Moa-nalo species all evolved from dabbling ducks. They were named Moa-nalo or "lost fowl" by Storrs L. Olson and Helen F. James because they were very different from any other anseriformes. Their massive bills had tooth like projections suitable for grazing and browsing. They were the largest herbivores on the Hawaiian islands and assumed the ecological role of the non-existent land tortoises.[4]

Geographic range: Islands of Maui and Moloka'i

Subfossil remains: Bishop Museum (Honolulu) and Smithsonian (Washington, D.C.)

On Moloka'i, a female Maui Nui Moa-nalo, *Thambetochen chauliodous*, protects its young from an attacking Hawaiian Eagle, *Haliaeetus albicilla*, now known to be a Hawaiian population of White-tailed Sea Eagle.[5] The plant in the foreground is the extremely rare *Schiedia lydgatei*, a member of the pink family (Caryophyllaceae) endemic to Moloka'i. (JULIAN P. HUME)

A pair of O'ahu Moa-nalo, *Thambetochen xanion,* feeding on the forest floor. The tooth-like projections in the jaw were used to strip tough-leaved plants like ferns. (JULIAN P. HUME)

O'ahu Moa-nalo

Thambetochen xanion

This flightless species was not as large as the Maui Nui Moa-nalo. Subfossil remains were found in sinkholes at Barbers Point and on the Mōkapu Peninsula in southeast O'ahu. The specific epithet, xanion, means comb in Greek referring to the bony, tooth-like projections on the jaws. The differentiation between these *T. xanion* and *T. chauliodous* seems rather slight given there has been no land connection between Maui Nui and O'ahu.[6]

Geographic range: Island of O'ahu

Subfossil remains: Bishop Museum (Honolulu)
and Smithsonian (Washington, D.C.)

Stumbling Moa-nalo

Ptaiochen pau

The Stumbling Moa-nalo lived in the mountainous areas on Maui and was different than *Thambetochen* because its jaw was both deeper and shorter and the tooth like projections were smaller.[7]

Geographic range: Island of Maui

Subfossil remains: Bishop Museum (Honolulu) and Smithsonian (Washington, D.C.)

Stumbling Moa-nalo, *Ptaiochen pau*, probably laid a single, large egg and nested on the ground. Behind the birds to the left is the critically endangered Rock Pelea, *Melicope balloui* (Rutaceae); in the center, the endangered *Remya mauiensis* (Asteraceae), and on the lower right the lobelioid *Delissea undulata*, now extinct on Maui but surviving on Hawai'i. (JULIAN P. HUME)

A group of Kaua'i Turtle-jawed Goose, *Chelychelynechen quassus,* avoid the heat under the endangered Hobdy's Wild-coffee, *Psychotria hobdyi* (Rubiaceae), (top) and critically endangered lobeliod, *Hibiscadelphis distans* (Malvaceae), (below). To the left of the geese, the fern, *Diellia manni* (Aspleniaceae), is down to a single individual, whereas the Serpent Woodfern, *Dryopteris crinalis* (Dryopteridaceae), to the right numbers around 47 individuals. Ferns would have been an important food source for the geese. (JULIAN P. HUME)

Kauaʻi Turtle-jawed Goose

Chelychelynechen quassus

The Kauaʻi Turtle-jawed Goose had the most developed turtle-like jaw of all the Moa-nalo species which it used for grazing the island's vegetation.[8]

Geographic range: Island of Kauaʻi

Subfossil remains: Bishop Muscum (Honolulu)
and Smithsonian (Washington, D.C.)

Kaua'i Mole Duck

Talpanas lippa

This flightless, almost blind species, was capable of foraging at night in closed canopy forest where there was abundant leaf litter. It is described as a duck that evolved into a Kiwi-like niche on prehistoric Kaua'i.[9]

Geographic range: Island of Kaua'i

Subfossil remains: Smithsonian (Washington, D.C.)

Kaua'i Mole Ducks, *Talpanas lippa,* forage on the forest floor for invertebrate prey with their broad, sensitive bills. The ancestors of these peculiar ducks may have been Stiff-tailed ducks, *Oxyura* sp. (JULIAN P. HUME)

Nocturnal Lesser Hawaiian Petrel, *Pterodroma jugabilis,* leave their nesting burrows from high in the mountains on O'ahu and Hawai'i to forage at sea. (JULIAN P. HUME)

Lesser Hawaiian Petrel

Pterodroma jugabilis

This small petrel's subfossil remains were found in a flooded sinkhole on O'ahu and in lava tubes in north Kona. It was similar in size to the Bonin Petrel and slightly larger than a Bulwer Petrel. The specific epithet, *jugabilis*, means "that may be joined" in Latin and is a reference to the unusual, nearly conjoined, supraorbital salt gland depressions on the cranium. The Lesser Hawaiian Petrel has no close relatives among living species of *Pterodroma*. Most of the subfossil bones of this species were found in an archaeological context, indicating that the birds had been used as food by ancient Hawaiians.[10]

Geographic range: Confirmed only from the Islands of O'ahu and Hawai'i but the former distribution of *P. jugabilis* will probably be known only with further paleontological work.

Subfossil remains: Bishop Museum (Honolulu)
and Smithsonian (Washington, D.C.)

Moloka'i Flightless Ibis

Apteribis glenos

This flightless ibis was described in 1976 from subfossil remains discovered in sand dunes at Mo'omomi by Joan Aidem. Later, additional remains of flightless ibis were found on Moloka'i and Maui. Storrs L. Olson and Helen F. James identified the Moloka'i Flightless Ibis and the Maui Flightless Ibis as separate species. The bill was long and very decurved with a narrow pointed tip. *Apteribis glenos* was an important prey item for the Hawaiian Eagle.[11]

Geographic range: Island of Moloka'i and Lāna'i

Subfossil remains: Bishop Museum (Honolulu)
and Smithsonian (Washington, D.C.)

Moloka'i Flightless Ibis, *Apteribis glenos*, search for invertebrate remains on the forest floor. This ibis probably occupied the same ecological niche as the New Zealand Kiwi, *Apteryx* sp. (JULIAN P. HUME)

A moment captured in time. A Maui Flightless Ibis, *Apteribis brevis,* falls into Auwahi Cave, Maui, where its remains were discovered millennia later by Storrs Olson and Helen James. The discovery of fossil remains by these paleontologists have shown that scores more taxa have disappeared from the Hawaiian Islands than was previously thought. (JULIAN P. HUME)

Maui Flightless Ibis

Apteribis brevis

This flightless ibis lived in the mountains on Maui. Its highly preserved remains were discovered in a lava tube. The nearly complete skeleton was collected on September 8, 1982 by Storrs L. Olson and Helen F. James. The Latin word *brevis* means short and was used because of this species short bill and small size. It is likely the ibis diet included land snails.[12]

Geographic range: Island of Maui

Subfossil remains: Bishop Museum (Honolulu)
and Smithsonian (Washington, D.C.)

Hawaiian Eagle

Haliaeetus albicilla ssp.

For over 100,000 years at least three of Hawai'i's main islands had resident populations of White-tailed Eagles. These majestic birds were the largest terrestrial predators in the archipelago. In 2014, phylogenetic analysis concluded the resident eagles were not clearly differentiated morphologically from mainland forms although they represent a distinct mtDNA (>3% divergent) lineage.[13]

Geographic range: Islands of Maui, Moloka'i and O'ahu

Subfossil remains: Washington, D.C.

On the coast of O'ahu, a Hawaiian Eagle, *Haliaeetus albicilla,* feeds its chick a Laysan Duck, *Anas laysanensis.* It is possible that this eagle nested in low vegetation or even on the ground. (JULIAN P. HUME)

A Hawaiian Harrier, *Circus dossenus*, plucks an ʻIʻiwi, *Vestiaria coccinea*, in koa forest on Oʻahu. (JULIAN P. HUME)

Hawaiian Harrier

Circus dossenus

The small Hawaiian Harrier was Accipter-like with short, broad wings. It inhabited the island's forests where it hunted small birds and insects. Subfossils were found in sinkholes at Barbers Point, Oʻahu and in sand dunes at Moʻomomi, Molokaʻi. The scientific name, *dossenus*, means clown or jester in Latin and refers to the initial confusion of this species generic placement.[14]

Geographic range: Islands of Oʻahu and Molokaʻi

Subfossil remains: Smithsonian (Washington, D.C.)

Ziegler's Crake

Zapornia ziegleri

This flightless species was described from subfossils found at Barbers Point, O'ahu that were collected on April 4, 1980 by Storrs L. Olson, Helen F. James, Aki Sinoto, and Eric Komori. The crake was named after Zoologist Alan C. Ziegler who participated in the important archaeological work at the 'Ewa sinkholes.[15] All of the flightless crake species in the Hawaiian Islands became extinct due to a combination of predation by introduced animals, over-hunting and deforestation.

Geographic range: Island of O'ahu

Subfossil remains: Smithsonian (Washington, D.C.)

A female Ziegler's Crake *Zapornia ziegleri* with her downy young on O'ahu. As in other flightless crakes, the number of young was probably two to three. All Hawaiian crakes were flightless, but evolved from an ancestral, volant *Zapornia* crake. Although often referred to as rails, the Hawaiian Rallidae are actually crakes, which differ from rails in having shorter bills.[16] (JULIAN P. HUME)

A tiny Moloka'i Crake, *Zapornia menehune*, searches for invertebrates beneath a Laysan Duck, *Anas laysanensis*. This crake may have measured just 10 cm in total length. (JULIAN P. HUME)

Moloka'i Crake

Zapornia menehune

The Moloka'i Crake is the smallest of all known Rallidae. It was described from subfossils found at Mo'omomi and 'Īlio Point dunes, Moloka'i. The scientific name *menehune* is a reference to the mythical leprechaun-like little people of the Hawaiian Islands.[17] It is possible the Moloka'i Crake survived into the early 20th century.

Geographic range: Island of Moloka'i

Subfossil remains: Smithsonian (Washington, D.C.)

Kepler's Crake

Zapornia keplerorum

This small flightless crake was described from subfossils found at Auwahi Cave, Maui in September 1982. It was larger than *Z. menehune* but smaller than any other known Hawaiian Rallidae except *Z. ziegleri*.[18]

Geographic range: Island of Maui

Subfossil remains: Smithsonian (Washington, D.C.)

Kepler's Crake, *Zapornia keplerorum,* scavenge around a basking Hawaiian Monk Seal, *Neomonachus schauinslandi,* on a Maui beach. No doubt Hawaiian crakes were omnivorous opportunists that took advantage of any food source. (JULIAN P. HUME)

A pair of Ralph's Crake *Zapornia ralphorum* scamper across weathered beach lava to escape from an O'ahu Hawk *Buteo* sp. The Hawaiian Hawk *Buteo solitarius* occurs only on the Big Island of Hawai'i, but it, or a closely related species, also inhabited O'ahu.[19] (JULIAN P. HUME)

Ralph's Crake

Zapornia ralphorum

Ralph's Crake was described from subfossils found at Barbers Point, O'ahu on July 10, 1981 by Storrs L. Olson, Helen F. James, James and Emilee Mead, Patrick McCoy and Virgil Meeker. It is larger than any other flightless Hawaiian Rallidae except *Z. severnsi*. The wing elements of the two O'ahu crake species were less reduced than their relatives on Maui.[20]

Geographic range: Island of O'ahu

Subfossil remains: Smithsonian (Washington, D.C.)

Severn's Crake

Zapornia severnsi

This large flightless crake species was described from subfossils found at Auwahi Cave, Maui. It is the most abundant crake found in the subfossil deposits on the island.[21]

Geographic range: Island of Maui

Subfossil remains: Smithsonian (Washington, D.C.)

A Severn's Crake, *Zapornia severnsi,* flees from a Greater Frigatebird, *Fregata minor,* into dune vegetation on Maui. (JULIAN P. HUME)

Kaua'i Stilt Owl

Grallistrix auceps

A new genus, *Grallistrix*, was created by Storrs L. Olson and Helen F. James for the species of long legged owls discovered from subfossil deposits on Kaua'i, O'ahu, Maui, and Moloka'i. Examination of fossil pellets of these owls proved they fed on birds. Their elongated legs and shortened wings were adaptations to help them catch avian prey. The Kaua'i Stilt Owl was described from subfossils found at Makawehi Dunes.[22]

Geographic range: Island of Kaua'i

Subfossil remains: Smithsonian (Washington, D.C.)

Mobbing of predators is common to many birds, and no doubt the Kaua'i Stilt Owl, *Grallistrix auceps,* was subject to the same behavior. Here a Kaua'i 'Ō'ō, *Moho braccatus,* (left); a pair of Kaua'i Nukupu'u, *Hemignathus lucidus hanapepe,* (top right); an 'Anianiau, *Loxops parvus,* (center right); and Kaua'i 'Amakihi, *Loxops stejnegeri,* mob a roosting Kaua'i Stilt Owl. (JULIAN P. HUME)

Previous page:
A Kaua'i Stilt Owl, *Grallistrix auceps,* attacks a group of Wahi Grosbeak, *Chloridops wahi,* and Kaua'i 'Ō'ō, *Moho braccatus,* on the Makawehi Dunes. The *Grallistrix* owls were probably diurnal hunters. (JULIAN P. HUME)

A Maui Stilt Owl, *Grallisttrix erdmani*, with a Maui ʻŌʻō, *Moho* sp. *Grallistrix* owls ranged from 30 to 43 cm in total length, had long legs and extremely well developed toes and long claws. These were adaptations for gripping and crushing their bird prey. (JULIAN P. HUME)

Maui Stilt Owl

Grallistrix erdmani

The Maui Stilt Owl was described from subfossil remains found at Owl Cave near Puʻu Makua, Maui. It was much smaller than *Grallistrix auceps* and *Grallistrix geleches*.[23]

Geographic range: Island of Maui

Subfossil remains: Smithsonian (Washington, D.C.)

O'ahu Stilt Owl

Grallistrix orion

The O'ahu Stilt Owl was described from subfossils discovered at Barbers Point, O'ahu in July 1981. It was smaller than the stilt owls found on Moloka'i and Kaua'i.[24]

Geographic range: Island of O'ahu

Subfossil remains: Smithsonian (Washington, D.C.)

A pair of Palila, *Loxioides bailleui,* feeding on the pods of Mamane, *Sophora chrysophylla* (Fabaceae), unaware of an attacking O'ahu Stilt Owl, *Grallistrix orion*. The Palila is today restricted to the higher elevations of Hawai'i, but the fossil record has shown that it once also occurred in coastal forests on O'ahu,[25] and perhaps on all of the main Hawaiian Islands. (JULIAN P. HUME)

A Moloka'i Stilt Owl, *Grallistrix geleches,* and chick await a meal on the Mo'omomi dunes. Owl pellets contain concentrations of well-preserved small bird bones, which have proved unprecedented in determining the original bird diversity of the Hawaiian Islands. (JULIAN P. HUME)

Moloka'i Stilt Owl

Grallistrix geleches

The long-legged Moloka'i Owl was described from a partial skeleton discovered on July 9, 1976 by Storrs L. Olson and Joan Aidem at Mo'omomi Dunes. The Hawaiian stilt owls roosted on sand dunes or in the low vegetation found in coastal areas.[26]

Geographic range: Island of Moloka'i

Subfossil remains: Smithsonian (Washington, D.C.)

Deep-billed Crow

Corvus impluviatus

Two never before described crow species were discovered by Storrs L. Olson, Helen F. James, Aki Sinoto and others at lowland fossil sites on O'ahu and Moloka'i. The well-preserved bones were collected under water in sinkholes at Barbers Point, O'ahu using scuba equipment. The Deep-billed Crow was a large crow with a high and deeply arched bill.[27]

Geographic range: Island of O'ahu

Subfossil remains: Smithsonian (Washington, D.C.)

Taking advantage of a fruiting Loulu Palm, *Pritchardia martii* (Arecaceae), a Deep-billed Crow, *Corvus impluviatus*, devours the ripe fruit, whereas O'ahu Palmcreepers, *Ciridops* sp., which were *Pritchardia* palm specialists[28], searched for invertebrate prey. (JULIAN P. HUME)

A pair of Robust Crows, *Corvus viriosus*, feed on the carcass of adult and chick Wedge-tailed Shearwater, *Puffinus pacificus,* on the Mo'omomi dunes, amongst Naupaka Kahakai, *Scaevola taccada* (Goodeniaceae), (left) and Akiaki grass, *Sporobolus virginicus* (Poaceae), (right). Moloka'i Crakes, *Zapornia menehune*, and a pair of Nihoa Finches, *Telespiza ultima,* share in the spoils. (JULIAN P. HUME)

Robust Crow

Corvus viriosus

The Robust Crow occurred on O'ahu where it lived with the Deep-billed Crow and also was found on Moloka'i. It was larger than the Hawaiian Crow, *Corvus hawaiiensis,* and had a long, straight bill. Regarding the three species of Corvus in Hawai'i, Helen James and Storrs L. Olson state, "The most obvious way to distinguish the three Hawaiian species of Corvus is by bill shape: *C. hawaiiensis* has a short, moderately arched bill; *C. impluviatus* has a high, impressively arched bill; and *C. viriosus* has a long, relatively straight bill." [29]

Geographic range: Islands of O'ahu and Moloka'i

Subfossil remains: Bishop Museum (Honolulu)
and Smithsonian (Washington, D.C.)

Narrow-billed Kioea

Chaetoptila sp.

The Narrow-billed Kioea formerly occurred on Maui together with the extinct *Chaetoptila aff. angustipluma*. The genus *Chaetoptila* was thought to be monotypic but the discovery of subfossils on Maui and O'ahu have shown this genus was originally widely distributed throughout the southeastern Hawaiian Islands.[30]

Geographic range: Island of Maui

Subfossil remains: Smithsonian (Washington, D.C.)

Narrow-billed Kioea, *Chaetoptila* sp., One of the extraordinary discoveries in recent years was the resolution of the true relationships of the Hawaiian Kioeas and 'Ō'ō, *Moho* sp. Once thought to be related to honeyeaters (Melipagidae), mDNA studies have shown that they are more closely related to Waxwings (Bombycillidae) and some other unusual passerines.[31] They have now been placed in their own family, Mohoidae. (JULIAN P. HUME)

Makawehi Finch *Telespiza persecutrix* (right) and Nihoa Finch *T. ultima* (center) await a chance to feed on an abandoned seabird egg on the dunes of O'ahu, pecked open by the larger Laysan Finch *T. cantans* (left). Laysan Finch also coexisted with Maui Nui Finch *T. ypsilon* and Nihoa Finch *T. ultima* on Moloka'i.[32] A young Ralph's Crake *Zapornia ralphorum* stalks insects attracted to the spoils. (JULIAN P. HUME)

Makawehi Finch

Telespiza persecutrix

This new species was described from subfossils collected from sinkholes at Barbers Point, Oʻahu on July 23, 1981 by Aki Sinoto and Patrick C. McCoy. Its former distribution also included the windward side of Oʻahu and the Makawehi Dunes area on Kauaʻi. The Makawehi Finch was closely related to the extant Laysan Finch and Nihoa Finch but it had a smaller bill.[33]

Geographic range: Islands of Oʻahu and Kauaʻi

Subfossil remains: Smithsonian (Washington, D.C.)

Maui Nui Finch

Telespiza ypsilon

Subfossils of this species were collected at ʻĪlio Point, Molokaʻi and Puʻu Naio Cave, Maui. The finch was fairly abundant in the harsh, arid environment of coastal Molokaʻi but was not restricted to these environments as the subfossils from Puʻu Naio Cave on Maui were found at 305-meter elevation in an area of more moderate climate. It was the most common prey item in the pellet deposits of the now extinct Molokaʻi Stilt Owl, *Grallistrix geleches*, at ʻĪlio Point.[34]

Geographic range: Islands of Maui and Molokaʻi

Subfossil remains: Smithsonian (Washington, D.C.)

A Maui Nui Finch, *Telespiza ypsilon*, plucks feather down from a molting Maui Nui Moa-nalo, *Thambetochen chauliodous*, for use as nesting material. (JULIAN P. HUME)

Giant Scimitar-billed Nukupu'u

Hemignathus vorpalis

This unusual species was described from a single partial skeleton discovered in a lava tube cave on the island of Hawai'i in 1992. It had a very large body compared to other species of *Hemignathus*.[35]

Geographic range: Island of Hawai'i

Subfossil remains: Bishop Museum (Honolulu)

A pair of Giant Scimitar-billed Nukupu'u, *Hemignathus vorpalis,* forage amongst leaf litter for invertebrate prey. It occurred in mesic montane forest on Hawai'i, dominated by 'ōhi'a, *Metrosideros polymorpha* (Myrtaceae), and Mamane, *Sophora chrysophylla* (Fabaceae), at 1,200 m above sea level, but probably utilised suitable lowland forest as well. (JULIAN P. HUME)

Comparison of Giant Scimitar-billed Nukupu'u, *Hemignathus vorpalis,* (below) compared with adult males of 'Akiapōlā'au, *H. wilsoni,* (above) and Kaua'i Nukupu'u, *H. lucidus hanepepe,* (middle). The dual-purpose bills of Nukupu'u and 'Akiapōlā'au are incredible enough, but the Giant Scimitar-billed Nukupu'u took this adaptation to an extreme. The purpose of the upper bill remains a mystery, but it may have been used to probe soft substrates or to move leaf litter while foraging. (JULIAN P. HUME)

Hoopoe-billed ʻAkialoa, *Akialoa upupirostris,* search for invertebrates on a rotting tree in the coastal forests of Kauaʻi. (JULIAN P. HUME)

Hoopoe-billed 'Akialoa

Akialoa upupirostris

This extinct long-billed species was collected at the Barbers Point site, O'ahu, and the Makawehi Dunes site on Kaua'i. It was similar to the other extinct 'Akialoa but had a shorter tongue. Helen F. James and Storrs L. Olson explain;

> "Remarks: The holotype and two paratypes are the only specimens that we have identified as A. upupirostris. We have not succeeded in associating a maxilla with this species: either it is morphologically inseparable from H. stejnegeri or we have not collected it yet. It might be questioned whether the fossils we have described as A. upupirostris are not in fact from the extinct Oahu Akialoa, A. lichtensteini. This is not the case, however, because the only two existing specimens of A. lichtensteini (ZMB 7918, holotype; ANSP 3360, previously misidentified as A. stejnegeri) have a pronounced lingual trough in the mandible (pers. observ.). We infer from the shape of the mandible that the tongue of A. upupirostris was shorter than the long, tubular organ in A. obscurus or A. stejnegeri. Lacking a trough on the anterior part of the mandible, the bill probably could not have accommodated such a long tongue. Although no other drepanidine lacks a well-developed trough on the dorsal surface of the mandible, this condition does occur in the hoopoes, woodhoopoes, and scythebills (Upupidae, Phoeniculidae, and Dendrocolaptinae: Campylorhamphus), with which A. upupirostris shares a convergent bill morphology (Olson and James 1982)."[36]

Geographic range: Islands of O'ahu and Kaua'i

Subfossil remains: Smithsonian (Washington, D.C.)

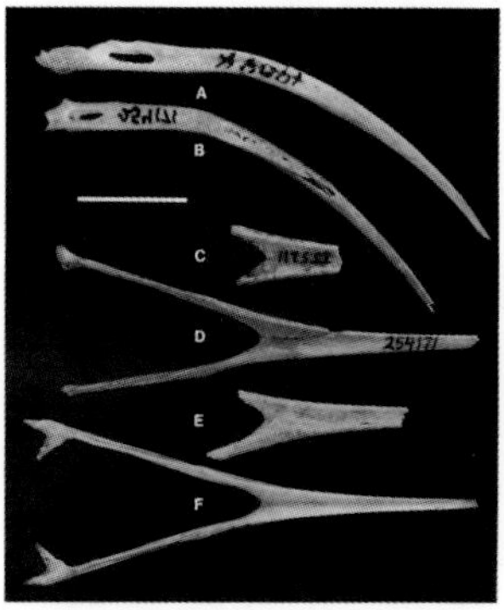

Mandibles of *Akialoa stejnegeri* (*H. procerus* auct.) and *A. upupirostris*, new species. **A** lateral and **F**, dorsal views of *A. stejnegeri* (USNM 19094); **B** lateral and **D** dorsal views of *A. upupirostris*, holotype (USNM 25471; lateral view is of the left side printed in reverse); **C** dorsal view of proximal portion of symphysis of *A. upupirostris* (USNM 255211); **E** dorsal view of symphysis lacking tip of *A. upupirostris* (USNM 255304). Scale = 1 cm. (COURTESY OF SMITHSONIAN INSTITUTION)

Kiwi Shovel-billed Finch

Vangulifer mirandus

A new genus *Vangulifer* or "little shovel bearer" was created for this species and *Vangulifer neophasis*. Both species had a unusual, blunt somewhat spatulate type bill. The Latin word *mirandus* means wonderful or strange. Subfossil remains of this species were found in Pu'u Naio Cave and at Lua Lepo on the island of Maui.[37]

Geographic range: Island of Maui

Subfossil remains: Smithsonian (Washington, D.C.)

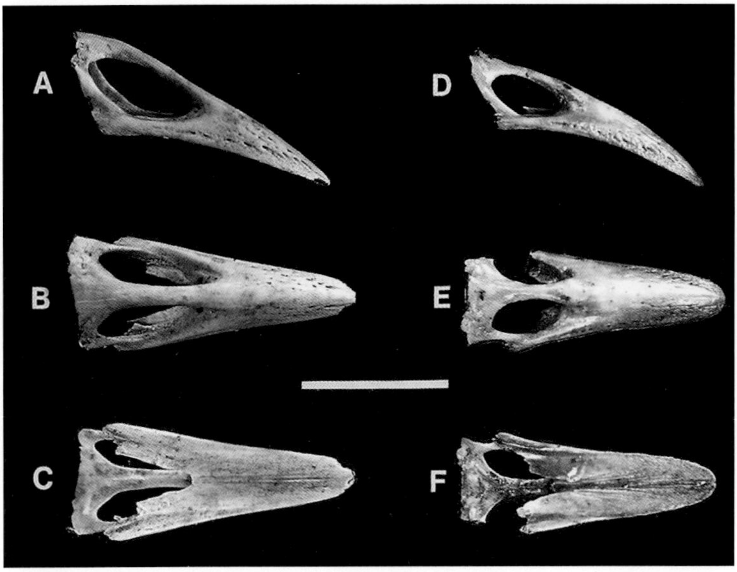

Maxillae of *Vangulifer*, new genus. **A** lateral, **B** dorsal and, **C** ventral views of *V. neophasis*, new species, holotype (USNM 44706 I); **D** lateral, **E** dorsal and **F** ventral views of *V. mirandus*, new species, holotype (USNM 445807). Scale = 1 cm. (COURTESY OF SMITHSONIAN INSTITUTION)

Previous page: A male Kiwi Shovel-billed Finch, *Vangulifer mirandus,* chases after a geometrid moth, *Scotorythra euryphaea* (Geometridae), in dense forest on Maui while a female looks on. The purpose of the unusual bill remains a mystery, but it may have been used to detect invertebrate prey in soft substrate or to hawk for insects like a flycatcher. (JULIAN P. HUME)

Pololei Shovel-billed Finch

Vangulifer neophasis

Subfossils of this new species were first collected on June 10, 1988 by Helen F. James at Pu'u Naio Cave on Maui. This species had a larger, straighter and deeper upper bill than *Vangulifer mirandus.*[38]

Geographic range: Island of Maui

Subfossil remains: Smithsonian (Washington, D.C.)

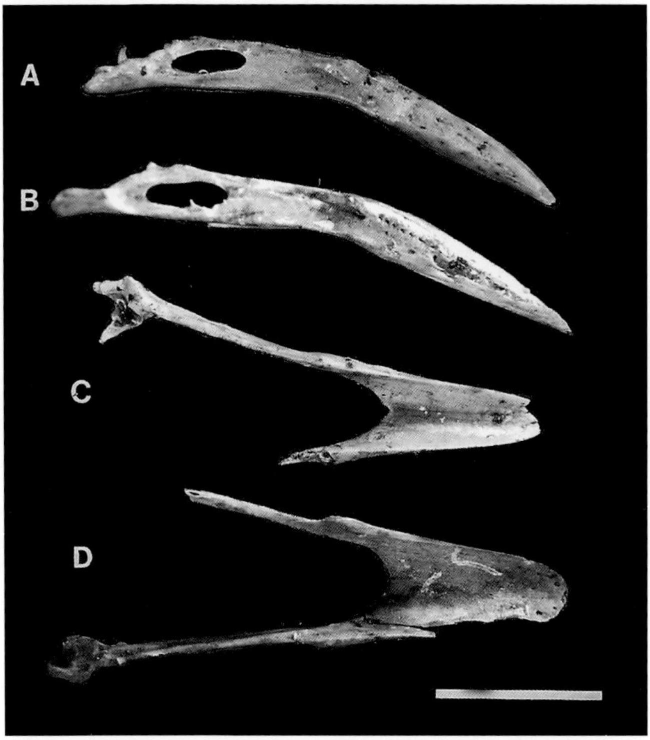

Mandibles of *Vangulifer*, new genus. **A** lateral and **C** dorsal views of *V. neophasis*, new species (USNM 445813; lateral view is of the left side printed in reverse); **B** lateral and **D** dorsal views of *V. rnirandus*, new species (USNM 445808). Scale = 1 cm. (COURTESY OF SMITHSONIAN INSTITUTION)

Previous page: A pair of Pololei Shovel-billed Finch, *Vangulifer neophasis*, forage near the ground for invertebrate prey in front of the extinct endemic mint, *Stenogyne haliakalae* (Lamiaceae). The mint was known only from the south slope of Haleakalā and has not been recorded since 1984. (JULIAN P. HUME)

Straight-billed Gaper

Aidemedia chascax

This new genus was named after Joan Aidem, one of the first collectors of bird subfossils in the Hawaiian islands. Remains of the Straight-billed Gaper were discovered at Barbers Point, O‘ahu in 1977. It is likely this species inserted its long, straight bill into soil, plants, flowers and fruit then used strong muscular action to spread the bill apart to obtain food.[39]

Geographic range: Island of O‘ahu

Subfossil remains: Smithsonian (Washington, D.C.)

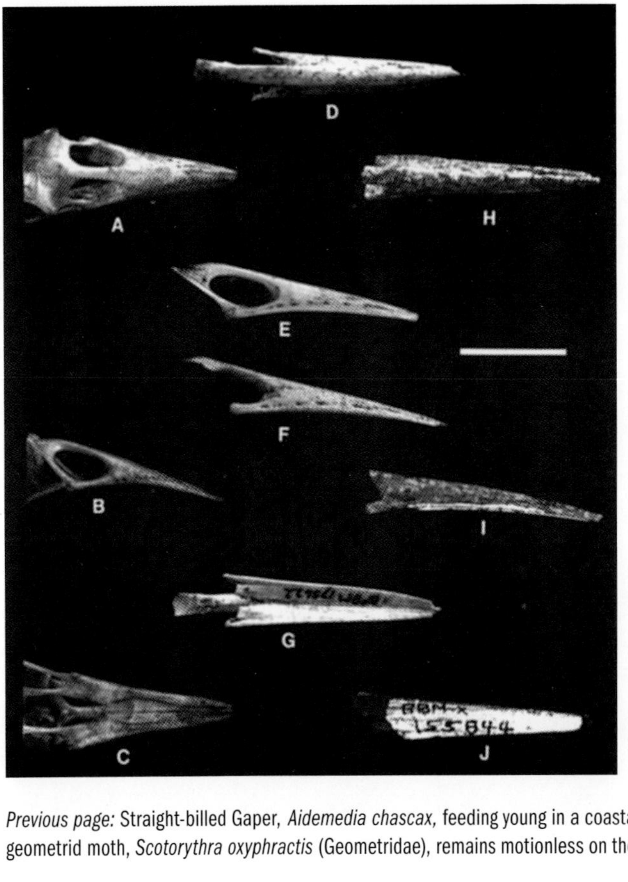

Maxillae of *Aidemedia*, new genus, and *Loxops sagittirostris*. **A** dorsal, **B** lateral and **C** ventral views of *L. sagittirostris* (AMNH 453236); **D** dorsal, **F** lateral and **G** ventral views of *A. lutetiae*, new species (BPBM 175622); **E** lateral view of *A. lutetiae* (USNM 445816); **H** dorsal, **I** lateral, and **J** ventral views of *A. chascax*, new species (BBM-X 155844). Scale = 1 cm. (COURTESY OF SMITHSONIAN INSTITUTION)

Previous page: Straight-billed Gaper, *Aidemedia chascax*, feeding young in a coastal forest on O‘ahu. A geometrid moth, *Scotorythra oxyphractis* (Geometridae), remains motionless on the bark (below). (JULIAN P. HUME)

Sickle-billed Gaper

Aidemedia zanclops

The Sickle-billed gaper (on right) also occurred on O'ahu and had a long, narrow, decurved bill. Subfossils were collected in July 1977 by Aki Sinoto and others at a site near Barbers Point. The scientific name *zanclops* is from Greek zanclon meaning, a sickle, plus ops, face, from the long, decurved mandible.[40]

Geographic range: Island of O'ahu

Subfossil remains: Smithsonian (Washington, D.C.)

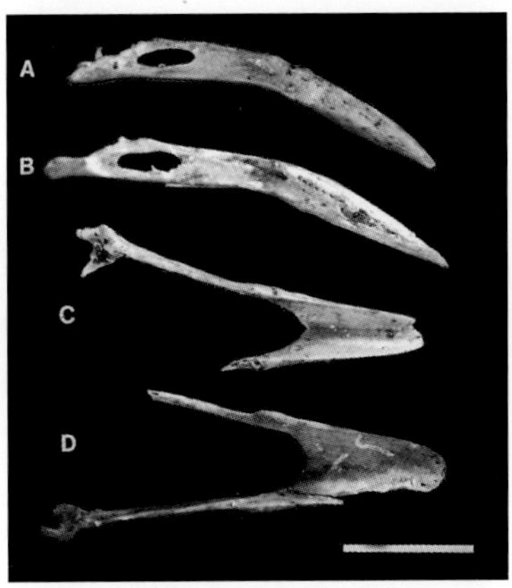

Maxillae of *Aidemedia*, new genus, and *Loxops sagittirostris* in dorsal view. **A** *A. zanclops*, new species, holotype (BBM-X 155160); **B** *A. chascax*, new species, holotype (BBM-X 155523); **C** *A. lutetiae*, new species, holotype (BBM-X 147441); **D** *L. sagittirostris* (AMNH 453236). Scale = 1 cm. (COURTESY OF SMITHSONIAN INSTITUTION)

The O'ahu gapers, Straight-billed Gaper, *Aidemedia chascax,* (left) and Sickle-billed Gaper, *A. zanclops,* (right) feed on the critically endangered giant lobelioid *Cyanea superba* (Campanulaceae) endemic to O'ahu. There is a possibility that these two gapers may belong to the same species and were sexually dimorphic, in which the male had a different bill shape to the female. (JULIAN P. HUME)

Maui Nui Gaper, *Aidemedia lutetiae,* forage in moss-covered bark searching for invertebrates. This species once occurred on Maui, Moloka'i, and possibly Lana'i as all of these islands were once connected creating the single large island of Maui Nui. (JULIAN P. HUME)

Maui Nui Gaper

Aidemedia lutetiae

Subfossils of this species were found at Moʻomomi Dunes on Molokaʻi and at Puʻu Naio Cave, Lua Lepo, and Puʻu Makua Cave on Maui. The Maui Nui Gaper had a short, very straight bill resembling that of a meadowlark.[41]

Geographic range: Islands of Molokaʻi and Maui

Subfossil remains: Smithsonian (Washington, D.C.)

Scissor-billed Koa Finch

Rhodacanthis forfex

Subfossils of this extinct species were collected by David Burney and others at the Makauwahi Cave site on Kaua'i in 1998.[42]

Geographic range: Islands of Maui and Kaua'i

Subfossil remains: Smithsonian (Washington, D.C.)

A male Scissor-billed Koa Finch, *Rhodacanthis forfex*, in a kūpuka on Kaua'i. The fossil record has shown that the koa finches were much more diverse than originally thought.[43] (JULIAN P. HUME)

A pair of Greater Koa Finches, *Rhodacanthis palmeri,* (lower pair; male above, female below) and Lesser Koa Finches, *Rhodacanthis flaviceps,* (upper pair; male above, female below) forage in a Koa, *Acacia koa* (Fabaceae), in the foothills of Haleakalā, Maui. Helen James and Storrs Olson found fossil bones of these two koa finches, *Rhodacanthis,* or something very closely related to them on Maui. (JULIAN P. HUME)

O'ahu Koa Finch

Rhodacanthis litotes

The O'ahu Koa Finch lived on the islands of O'ahu and Maui before it became extinct after the colonization of the Hawaiian Islands by Polynesian settlers.[44]

Geographic range: Islands of O'ahu and Maui

Subfossil remains: Bishop Museum (Honolulu)
and Smithsonian (Washington, D.C.)

O'ahu Koa Finch, *Rhodacanthis litotes*, in a Koa, *Acacia koa* (Fabaceae). The historically extinct koa finches of Hawai'i were known to feed almost exclusively on koa pods, so it is likely that the fossil species did as well. (JULIAN P. HUME)

Wahi Grosbeak

Chloridops wahi

This species was similar to the extinct Kona Grosbeak *(Chloridops kona)* but had a smaller and shorter bill. Subfossils were collected in August 1977 by Aki Sinoto at Barbers Point, O'ahu. The scientific name "wahi" is Hawaiian for to cleave or break in pieces; a reference to the function of the bill to crack hard seeds.[45]

Geographic range: Islands of O'ahu and Maui

Subfossil remains: Bishop Museum (Honolulu) and Smithsonian (Washington, D.C.)

King Kong Grosbeak

Chloridops regiskongi

The King Kong Grosbeak had the largest bill known of all the Hawaiian finches. Subfossils were collected on August 14, 1981 by Aki Sinoto and others from sinkholes located near Barbers Point, O'ahu.

The following quotation from *Descriptions of thirty-two new species of birds from the Hawaiian Islands: Part 2. Passeriformes* explains the origin of the interesting name:

> In a newspaper article, Olson was once quoted as saying this species was "a giant, gargantuan, a King Kong finch" (Benson 1977), an appellation that would never have occurred to him, this being a typical example of the liberties taken with quotation marks by the print media. Nevertheless, because of its ridiculousness, the name naturally stuck and we found ourselves referring jocularly to this species as the "King Kong finch" ever afterwards.[46]

Geographic range: Island of O'ahu

Subfossil remains: Bishop Museum (Honolulu) and Smithsonian (Washington, D.C.)

Previous page: Wahi Grosbeak, *Chloridops wahi,* (above) and the King Kong Finch, *C. regiskongi,* (below) feeding on the fruit of Sandalwood, *Santalum freycinetianum* (Santalaceae), on O'ahu. The King Kong Finch was the largest of the finch-like Hawaiian finches and had evolved a huge, seed crushing bill. (JULIAN P. HUME)

Mauka Grosbeak, *Orthiospiza howarthi,* feeding its young in the mountain forests of Maui. As the bill of this species differed from *Chloridops* and *Rhodacanthis* in being comparatively short and broad, it may have had a different feeding ecology. (JULIAN P. HUME)

Mauka Grosbeak

Orthiospiza howarthi

This species was described from subfossils collected at Lua Manu, Puʻu Makua Cave, and Auwahi Cave on Maui. It was comparable in size to the extinct Greater Koa Finch, *Rhodacanthis palmeri*. It is possible the Mauka Grosbeak was restricted to montane habitats.[47]

Geographic range: Island of Maui

Subfossil remains: Bishop Museum (Honolulu) and Smithsonian (Washington, D.C.)

Pila's Palila

Loxioides kikuchi

This new species of Hawaiian finch was described from subfossils found at Makauwahi Cave on Kaua'i. It was larger than the extant Palila, *Loxioides bailleui*, and lacked a hooked bill. Both species of Palila once coexisted in the coastal areas of Kaua'i.[48]

Geographic range: Island of Kaua'i

Subfossil remains: Smithsonian (Washington, D.C.)

Pila's Palila, *Loxioides kikuchi*, in a coastal forest on Kaua'i. As Pila's Palila differed from the Palila *Loxioides bailleui* in its bill structure, it may have fed on different types of food; thus both species could coexist in similar habitats. (JULIAN P. HUME)

A pair of Cone-billed Finch, *Xestospiza conica,* (right) forage on *Kanaloa kahoolawensis* (Fabaceae), a legume known from only two wild individuals on Kaho'olawe; however, it once occurred on most of the large islands. A Kaua'i Thrush, *Myadestes myadestinus,* perches alongside. (JULIAN P. HUME)

Cone-billed Finch

Xestospiza conica

Storrs L. Olson and Helen F. James created the new genus, *Xestopiza*, for two extinct Hawaiian finch species that had long, narrow, cone-shaped bills instead of arched bills. This species' remains were found in pellets of the extinct Kaua'i Stilt Owl, *Grallistrix auceps.*[49]

Geographic range: Island of Kaua'i

Subfossil remains: Smithsonian (Washington, D.C.)

Ridge-billed Finch

Xestospiza fastigialis

The Ridge-billed Finch had a resemblance to cowbirds (Molothrus). It was the most widely distributed of all the new species described from subfossils.[50]

Geographic range: Islands of O'ahu, Moloka'i, and Maui

Subfossil remains: Bishop Museum (Honolulu)
and Smithsonian (Washington, D.C.)

A Ridge-billed Finch, *Xestospiza fastigialis*, preying on the critically endangered Crimson Hawaiian Damselfly, *Megalagrion leptodemas* (Coenagrionidea), in a forested drainage area on O'ahu. (JULIAN P. HUME)

Kaua'i Palmcreeper

Ciridops tenax

This extinct species was described from subfossils found at Makawehi Dunes on Kaua'i. The Kaua'i Palmcreeper was smaller than the extinct 'Ula-'ai-hawane, *Ciridops anna.* Similar to other newly described Hawaiian species, the subfossil remains were discovered inside pellets of the Kaua'i Stilt-legged Owl, *Grallistrix auceps.*[51]

Geographic range: Island of Kaua'i

Subfossil remains: Bishop Museum (Honolulu) and Smithsonian (Washington, D.C.)

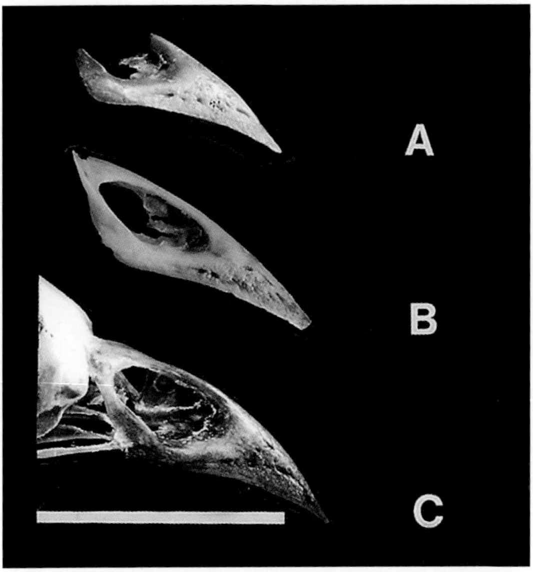

Maxillae of Ciridops in lateral view. **A** *Ciridops* sp., Oahu (BBM-X 155727); **B** *C. tenax*, new species, holotype (USNM 254913); **C** *C. anna* (MCZ 10995). Scale = 1 cm. (COURTESY OF SMITHSONIAN INSTITUTION)

Previous page: Two male Kaua'i Palmcreepers, *Ciridops tenax,* (above) and a female (below) feed in the crown of a Loulu Palm, *Pritchardia* sp. The palmcreepers had evolved robust legs and feet for foraging in the matted mass of palm inflorescences in order to obtain invertebrates.[52] (JULIAN P. HUME)

JP Hume

Extinctions Since 1778

Kona Finch, *Chloridops kona,* feeding chicks with looper caterpillars. (JULIAN P. HUME)

Laysan Crake, *Zapornia palmeri*, search for food around Laysan Albatross, *Phoebastria immutabilis*, and alongside the endangered pea, 'Ōhai *Sesbania tomentosa* (Fabaceae), left. (JULIAN P. HUME)

Laysan Crake

Zapornia palmeri

Description:

Adult male. Top of the head and entire upper surface with tail pale sandy brown; feathers slate-coloured and whitish along the shaft at base, and with large deep blackish-brown central patches on the exposed parts; the feathers of the back are mostly mottled with white, but in some specimens the white is scarcely perceptible; a line over the eye, sides of the head and neck, chin, throat, breast, and abdomen pale slaty grey to ashy; feathers on the sides of the body of the colour of the upper surface, each with several white spots, faintly outlined with black. Wings very small and rounded, first primary about a third of an inch shorter than the second; second, third, and fourth about equal and longest; outer webs sandy brown with very pale margins, inner webs darker and more smoky brown. Bill light olive-green, darker and inclining to purplish on the tip and culmen; iris ruby-red; eyelid pale grey-green; tarsus and feet light olive-grey-green.[1]

The sexes are similar, only that the females seem to be a little paler. Young birds have the entire under surface pale buff. The nestlings have a yellow bill and are covered with black down.[2]

Measurements:

Total length about 6 inches (but the bird is capable of extending its neck considerably, adding much to its entire length), culmen 0.65 to 0.7, wing 2.2 to 2.3, tarsus 0.9 to 1, middle toe with claw 1.2 to 1.3, hind toe with claw 0.4. The tail is very inconspicuous and only 0.8 to 0.9 inch in length, the tail-feathers only a little stiffer than the coverts and so much like them that it is almost impossible to make out their exact number.[3]

—Walter Rothschild, 1900

Geographic range: Laysan Island

Specimens in museums: 250

Historical information: First observed 1828. Last observed July 1944.

Reports and observations:

A kind of *Fowl*, about as large as a Ptarmigan; mixed grey and brown; running on the ground, singly, but at the same time rather numerous on Moller* and Lisiansky; very rapid and rather shy.[4]

—C. Isenbeck, 1828

**(An early name for Laysan Island)*

June 16—At last we reached Laysan, having sighted it just after daylight. From what I can see of the island I am in hopes that I shall be able to make up a little of the lost time, for, to my great delight, the island appears quite alive with birds, especially the Rail, which, although I had been told was tolerably plentiful, to my intense astonishment literally covers the island. Everywhere one walks this little creature hurries out of your way.[5]

A unique photograph taken around 1939-40 by Thomas M. Blackman of a living Laysan Crake *Zapornia palmeri* on Midway Atoll.[6] An introduction of 83 birds to the atoll in 1913 was to safeguard the population because of the impending denudation of Laysan's vegetation by rabbits. The last crake perished in May 1923 on Laysan. After rabbit removal, Laysan's flora gradually recovered, so contingency plans ensued to return crakes to Laysan. However, despite numerous efforts to do so by conservationists, misguided bureaucracy by the Hawaii Board of Agriculture and Forestry (HBAF) prevented transportation.[7] Subsequently, rats invaded Midway Atoll during World War II, and by July 1944, the crake was gone. The loss of this unique bird is another sad episode in the history of human-caused extinctions, especially as in this case it was so easily preventable.

June 19—Again the day was spent in collecting and preparing birds. While out this morning both my assistant and I saw a little Rail break and eat an egg. We had disturbed from a nest a Noddy *(Anous)*, immediately the Rail ran up and began to strike at the eggshell with its bill, but the egg being large and hard he was quite a long time before making a hole. The Rail would jump high into the air and come down with all its force on the egg, until it accomplished the task, which once done the egg was soon emptied. By this time the Tern came back and gave chase but in vain. At the beginning I ran after the Rails with the net, but soon found it was much better to put the net edgeways on the ground, when the very inquisitive Rails would run up to look at it, and then were easily caught.[8]

June 24—My assistant and I went out again this morning searching for eggs of the land-birds, and I am pleased to say we found two nests of the Rail. One nest contained three eggs, the other an egg and small chick. Since bringing the eggs home, the one found with the chick has hatched into a young Rail, which I have put in spirits. I now have Rails of every size from the egg upwards.[9]

—Henry C. Palmer, 1891

The Laysan Rail is a wide-awake, inquisitive little creature, with an insatiable thirst for first-hand knowledge. It is one of the most naïve, unsophisticated, and wholly unsuspicious birds in the whole avian catalogue. At times it is confiding and familiar in deportment, yet at others holds aloof with some show of reserve. It will occasionally hide behind a bunch of grass, as if afraid, and then suddenly come forth with entire change of demeanor and examine the intruder with critical eye.[10]

Laysan Crake in a nest. (WALTER K. FISHER– ALL RIGHTS RESERVED. IMAGE ARCHIVES. DENVER MUSEUM OF NATURE & SCIENCE)

The rails are everywhere on Laysan in great numbers. Nearly every bunch of grass seemed to harbor a pair. They probably have no enemies of any importance, and the only check to their increase is space and food supply.[11]

In strolling through the brush we could hear them calling on all sides. Their 'song' is a plaintive, high-keyed little rattle, which resembles remotely an alarm clock with a muffled bell or pebbles ricocheting on a glass roof.[12]

—Walter K. Fisher, 1902

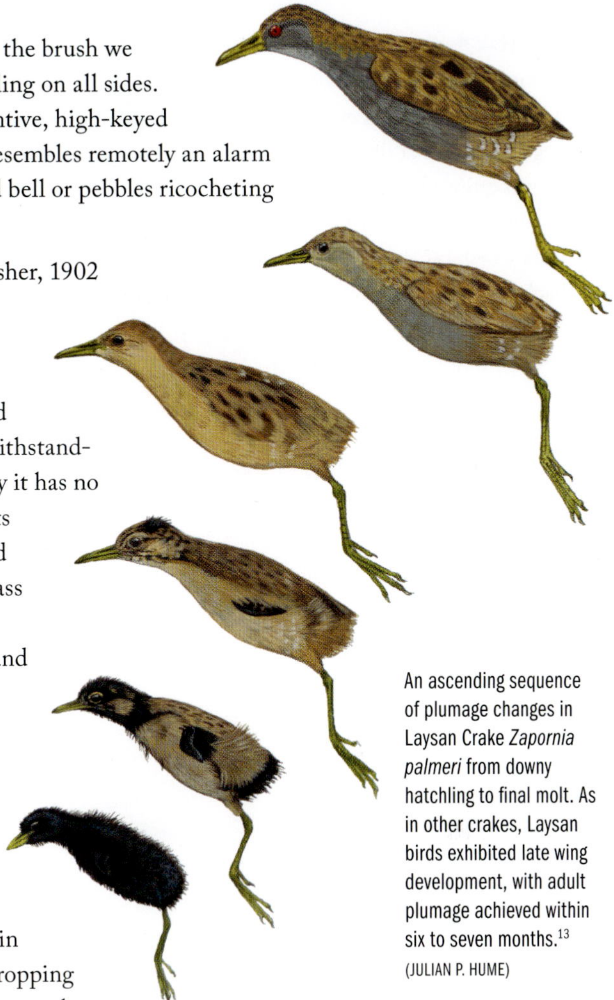

An ascending sequence of plumage changes in Laysan Crake *Zapornia palmeri* from downy hatchling to final molt. As in other crakes, Laysan birds exhibited late wing development, with adult plumage achieved within six to seven months.[13] (JULIAN P. HUME)

The Laysan Rail is one of the most interesting birds found on the island. Notwithstanding its inability to fly it has no trouble in evading its pursuers. It runs and dodges from one grass tussock to another, down a petrel hole and out again before one can locate it. One of the most laughable things imaginable is a man pursuing one of these bits of bird life, net in hand, continually dropping waist-deep down among the burrowing petrels. It was with much difficulty that we secured the specimens we needed for our collection. They were everywhere fairly abundant on all parts of the island excepting the beaches. They even visited our kitchen. Their favorite nesting place is among the thick mats of juncus along the margin of the lagoon. We found many nests but no eggs. On the 4th day of June we discovered two coal-black chicks with yellow beaks that gave vent to much noise. There are about 2,000 rails on the island.[14]

Mr. Max Schlemmer, the former manager of the guano company on Laysan, introduced domestic rabbits, Belgian hares and English hares into the island at different times, and left them there to breed, intending to start a rabbit canning business. He could not give the exact dates but thinks the first were imported eight or nine years ago or about 1903…Rabbits have killed many bushes, but have not caused so great devestation as one would expect from their numbers…It is my opinion that a well-organized party with small repeating rifles, say 22 caliber, could entirely exterminate these pests.[15]

Laysan Crake. (ALFRED M. BAILEY–ALL RIGHTS RESERVED. IMAGE ARCHIVES. DENVER MUSEUM OF NATURE & SCIENCE)

—Homer R. Dill, 1911

Yes, the birds are here at least in so far as the common sea fowl go but the striking thing is the VEGETATION—or utter lack of it: rather! Verily, the damned rabbits have done their worst. As far as I can see with the glasses and from our hurried trip down the island there is not a living bush or twig or spear of grass left on the whole island outside of the two poor cocoanut trees and the 3 bushes near the house that Sula piscator has preempted—and two of the latter are leafless! Apparently the swarming host of rabbits has gone the limit. The once green island is today a desolate sand waste that might as well be a salt sink in one of our barest sand deserts. In fact the salt lagoon which occupies the whole center of the island, and to which all the fringing ridges of sand slope down, gives the simile striking aptitude. In my wildest pessimism I had not feared such utter extirpation of every living plant. I had hoped that the rabbits had wrought their own destruction in the general ruin, but those who went east and north report two small patches of close-cropped "pickle" weed (*Sesuvium* sp?) that still support a remnant of the cursed host. The sailors killed them with stones and Reno with a rifle. In all probably 50 were accounted for.[16]

Schleimmer came up holding a Laysan Rail in his hand. To get all the footage possible we turned him loose in front of the high speed. Meantime I had held alive and unhurt in my hand one of the two Laysan Rail we know are left on the island, noted his red iris and green (basally) bill and depauperate wings, developed plumage.[17]

—Donald R. Dickey, 1923

(Laysan Crakes were never observed again on Laysan Island)

Eastern Island of Midway Atoll was probably the last home of this rail, as E. L. Caum and Sergeant Lewis W. Walker, U.S.M.C., saw two there in July, 1944. The main factor in the extermination of the rail was the overrunning of both islets with rats, which escaped to Midway from ships in 1943. Had the birds survived the extreme rat infestation of 1943 and 1944, they might still find cover and continue reproducing now that the numbers of rats are reduced.[18]

—Harvey I. Fisher and Paul H. Baldwin, 1946

Video of the Laysan Crake can be found at this link:
https://www.youtube.com/watch?v=0dhDjSF0O1c

F. W. Frohawk del. et lith

West, Newman imp.

PENNULA WILSONI

Western Hawaiian Crake

Zapornia sandwichensis sandwichensis

Description:

Size small. Bill dusky ash-colour: general colour of the plumage pale ferruginous; the feathers on the upper parts darkest in the middle: tail short, hid by the upper coverts: legs dusky flesh-colour[19]

—John Latham, 1785

Geographic range: Island of Hawai'i, Western side.

Specimens in museums: 2

Historical information: First specimen collected 1779. Last observed 1779.

"Two specimens, now in museums in Leiden and Vienna, were collected near Kealakekua during the third visit by James Cook (Stresemann 1950, Medway 1981), probably on 26 Jan 1779, when a landing party purchased "two small fowl" from some taro farmers (Ellis 1782:92)"[20]

Reports and observations:

...A Rail with very short wings and no tail, which on that account we named, *Rallus ecaudatus*...[21]

—Captain Charles Clerke, 1779

F. W. Frohawk del. et lith

West, Newman imp.

PENNULA SANDVICENSIS

Fac-simile of Ellis's drawing dated 1779.

F. W. Frohawk del. et lith

West, Newman imp.

PENNULA ECAUDATA

Eastern Hawaiian Crake

Zapornia sandwichensis millsi

Description:

Upperside dark chocolate-brown, a little paler on the head, darkest and slightly mottled by the faint appearance of the almost black centres of the feathers on the rump and upper tail-coverts. Wing-coverts like the back. Quills blackish, the edges chocolate-brown. Rectrices blackish, with chocolate-brown tips. Sides of the head with a ashy tinge. Underside lighter. Chin palest, whitish brown. Throat and chest vinous-brown, shading off into dusky chocolate-brown on the abdomen and under tail-coverts. Some of the feathers on the sides of the rump with indications of rufous-buff cross-bars. Under wing-coverts dark chocolate-brown.

Dimensions:

Total length about 5.5 inches, wing 2.7, tail 0.55, metatarsus 1.1, middle toe with claw 1.5, hind toe with claw 0.55, culmen 0.75[22]

—Walter Rothschild, 1900

Geographic range: Island of Hawai'i, Eastern side.

Specimens in museums: 5

Historical information: First specimen collected 1859.
Last observed 1884.

Reports and observations:

The inference would seem to be that at the date of Mr. Pease's letter (20th November, 1861) Mr. Mills possessed only one specimen, and that the other four which I myself saw were obtained subsequently. Two of them have since passed into Mr. Rothschild's collection, and the remaining two are still in that of Mr. Bishop. No further examples have been secured, though it is doubtful whether any extended search has been made.[23]

J. G. Keulemans del. et. lith — Mintern Bros imp.

PENNULA SANDWICHENSIS (GM)
IN THE LEYDEN MUSEUM

PENNULA MILLSI, DOLE.
IN THE TRING MUSEUM

Moreover, Hawelu told me that the mail-carrier had seen the bird cross his path within the last three years; on the same authority, the Moho outruns any dog possessed by the natives, and it is possible to track it by its cry—a whirring sound resembling the rising of a bevy of Quail, while its nest is made on the ground. The five specimens were all procured by Hawelu in the scrub-covered lava-flats about five miles south of the Volcano House, but more information than this I could not obtain, and my intelligent informant is now a leper on Molokai.[24]

I may add that the late King Kalakaua was most anxious to procure specimens of the subject of the present notice, and had for some years before my visit offered the natives a large reward for them. Olaa used to be a noted locality in olden times for bird-catching, and his late Majesty, through his Chamberlain, Mr. C. P. Jaukea, gave me a written permission to shoot specimens of the Mamo *(Drepanis pacifica)* and O-o *(Acrulocercus nobilis)* there, believing that I should meet with both of them as well as the Moho; he afterwards expressed great disappointment at my failing in my object.[25]

—Scott B. Wilson, 1899

There can be no doubt that this rail is now quite extinct. Neither Mr. Wilson nor Mr. Perkins found it. Moreover, I sent out to Palmer a dog which was specially trained for the purpose of getting Rails, and he stayed a good time at Olaa without seeing a sign of Rails.[26]

—Walter Rothschild, 1900

Little is known about the moho. Its nest was seen by several persons in the grass but there is no description of the nest, eggs or young. The Hawaiian chiefs used to find sport in shooting at this bird with bow and arrow.[27]

—George C. Munro, 1944

Illustrator unknown

PENNULA SANDWICHENSIS

F. W. Frohawk del. et lith

West, Newman imp.

ACRULOCERCUS BRACCATUS

Kaua'i 'Ō'ō

Moho braccatus

Description:

Adult male. Head black, streaked with a few longitudinal lines of white; rest of the upper surface slaty brown, brightening into russet on the rump and flanks; throat and breast black, each feather barred with white; rest of the under surface dull slaty brown, while the centres of the feathers being grey give it a streaked appearance; wings and tail black, the central pair of feathers of the latter much exceeding the rest in length; axillary tufts (little developed) of a pale greyish buff; edge of the wing pure white; tibiae rich golden yellow; irides light yellow; bill and feet black.[28]

Adult female. Similar to the male, but with the feathers on the throat much more extensively barred with white, which gives the bird the appearance of having a well defined whitish patch on the throat and upper part of the breast.[29]

Measurements:

Total length 7.75 inches, wing from carpal joint 3.90, culmen 1.5, tarsus 1.5, tail 3.5.[30]

—Scott B. Wilson, 1899

Geographic range: Island of Kaua'i

Specimens in museums: 111

Historical information: First specimen collected 1835. Last heard 1987.

Reports and observations:

Another bird is called Oo by the natives; it is *Certhia pacifica* of Latham, and is found on the island of Kauai, one of the same group. It also has tufts of yellow feathers which have been collected for the same purpose in former days; theirs are on the thighs, not on the sides as in the genuine Oo; the feathers are smaller, much inferior in beautiful texture, and are no longer collected; both species are black. We killed specimens at Hanalei, a department of the Island of Kauai, where they are found in the woody districts on the mountains.[31]

—Titian R. Peale, 1848

J. G. Keulemans del. et. lith Mintern Bros imp.
MOHO BRACCATUS, CASS. 1 AD. 2 JUV.

Kaua'i ʻŌʻō, *Moho braccatus*.
(AUCKLAND MUSEUM NEW ZEALAND)

With regard to the call-note—a kind of chuck—it is noteworthy that it is somewhat similar to that of the larger O-O, though in a higher key; the bird has also a sweet song, some of its notes possessing a bell-like clearness.[32]

It is common in the woods by which Makaweli, the lovely mountain residence of the Sinclair family, is surrounded, where in the early morning its dulcet tones may be heard to perfection, blended with those of its forest companions…[33]

This bird is not nearly so wary as *A.nobilis*, but is very clever in concealing itself among the thick foliage, thereby rendering observation difficult.[34]

—Scott B. Wilson, 1899

When I sat down under the tree and imitated its cry, half a dozen came right close to me; then seeing my dog move gave out shrill squeaking notes which I think were alarm notes, for instantly others answered, and though not much frightened, they were now more cautious in coming close to me.[35]

—Henry C. Palmer, 1891

Wilson says it seems to be found at all elevations throughout the forests, and is called "O-o A-a," the dwarf O-o by the natives.[36]

Palmer found this bird common on Kauai. He saw it more often than his congeners on the trunks of trees, especially Ohias, pecking in crevices of the bark. It was very numerous on flowering Ohias. On one single tree of that kind Palmer saw more than once about a dozen at the same time.[37]

—Walter Rothschild, 1900

In 1891 it was a common bird over all the Kauai forests. Its notes could be heard from near sea-level in the valleys on the north side to near the top of Mt. Waialeale at over 4,000 feet elevation. It was still not uncommon when I left Kauai in 1899, but on four visits to the Kaua'i forests between 1928 and 1936 I failed to hear or see it.[38]

—George C. Munro, 1944

Kaua'i 'Ō'ō, *Moho braccatus*, in Alaka'i Swamp, Kaua'i, Hawai'i. (© ROBERT SHALLENBERGER)

It may well have been seen or heard by others, too, during the last half century, but we were able to establish its current existence by collecting a specimen on July 21 (probably the first since 1897) and by closely observing or hearing at least 12 other individuals.[39]

—Frank Richardson and John Bowles, 1964

By 1980 and during the HFBS in 1981 only a single pair could be found attending a nest cavity along Halepa'akai Stream at about 1300 m elevation (Hart 1978; Scott and Kepler 1985; Scott et al. 1985, 1986; Sykes et al. 2000). The female was last reported in May 1981 and the male continued to be observed "going through nest-building ritual in spring, and calling over and over for a mate that may never come" (AB 37:916) until last heard on 29 Apr 1987 (E 47:94, AB 41:492). There have subsequently been no substantiated observations, despite numerous searches (e.g., Reynolds and Snetsinger 2001, Foster et al. 2004, USFWS 2006, Paxton et al. 2016; E 53:49), and we presume the Kaua'i 'O'o is now extinct. Based on Poisson analyses of persistence probabilities using confirmed and unconfirmed records, Elphick et al. (2009) estimated that the Kaua'i 'O'o went extinct in 1985-1991, with upper limits of 2000-2007; see also Reynolds and Snetsinger 2001 and Scott et al. 2008.[40]

—R. L. Pyle and P. Pyle, 2017

*HFBS=Hawai'i Forest Bird Survey

F. W. Frohawk del. et lith. West, Newman imp.

ACRULOCERCUS APICALIS

O‘ahu ‘Ō‘ō

Moho apicalis

Description:

Adult male. Black; brownish and with distinct pale shaft-lines on the back and rump. Rectrices with white tips for about 0.6 to 0.1 in., only the central pair without white tips; the central pair ending in a large narrow point, which is twisted upwards. A broad line along the sides of the breast and flanks; vent and under tail-coverts yellow. Under wing-coverts, with the exception of the outer ones near the bend of the wing, white. Bill and feet blackish.[41]

The *adult female* is similar to the male, but smaller.[42]

Measurements:

Wing 4.65 inches, tail 6.2, culmen 1.8, tarsus 1.5.[43]

—Walter Rothschild, 1900

Geographic range: Island of O‘ahu.

Specimens in museums: Ten.

Historical information: First specimen collected 1786-1787. Last observation 1837.

L. from 12 to 14 inch. Bill black, slightly curved & sharp pointed, VA long. Nostrils linear, at the base covered with a hard membrane. Tongue very long & tubular, divided half way into two threads, each of which is again divided halfway into two more. The extremities of all four are divided into minute filaments. Whole of the bird a deep shining black, except a line of yellow feathers extending on each side from under the wings to the vent. These are the feathers esteemed so valuable by the natives for their softness and scarcity & the royal clokes are principally made from them…Their note is a harsh chirp of two or three different tones. They feed principally on the nectar of the Bowers of the *Eugenia Malaccensis*.[44]

I kept some alive which the natives brought me, tho' almost entirely destitute of feathers, for three of four weeks, feeding them principally upon sugar & water. They took flies, however, which came into their cage, with great quickness & adroitness. They are apparently a strong & hardy bird, but the cold would probably besides the nature of their food, hinder them from being taken to England. They are now very scarce in all the islands. I did not see even one in the different excursions I made & the natives asked a high price for the very few they brought to me & almost the whole of these were destitute of feathers. I preserved only one tolerable specimen the whole time I was upon the islands—& even from that some of the yellow feathers had been plucked out.[45]

—Alexander J. Bloxam, 1824

Several days ago Mr. Deppe and myself visited Nuano [Nuuanu] valley, where we hired a native house, in which we are now living. Our object has been to procure birds, plants &c. and we have been so far very successful. I have already prepared about eighty birds that I have procured here.[46]

They are also very successful bird-catchers. In the island of Oahu they procure the gum of a tree which they call "Tu-tui," and make of it a tenacious paste by moistening it with water. They smear little sticks with this paste and plunge them into the large pods of the banana, which contain a sweet juice of which the bird is extremely fond. The bird alights upon the gummed stick, and his feet are in an instant so firmly glued to it, that he cannot escape. By this mode, dozens of beautiful birds were brought to me almost daily, all alive and uninjured.[47]

—John K. Townsend, 1837

A male (above) and female (below) O'ahu ʻŌʻō *Moho apicalis* held at the Museum für Naturkunde, Berlin. The German naturalist, Ferdinand Deppe (1794-1861), obtained this pair and one other in January 1837 in the Nu'uanu Valley. These are the last known specimens. Note the striking size difference between the sexes (sexual dimorphism). (JULIAN P. HUME)

J. G. Keulemans del. et. lith Mintern Bros imp.

MOHO APICALIS, GOULD

(1) Male, (2) Female

This, the o-o of Oahu no doubt was formerly common, but none of the recent explorers have found it, and the presumption is that it is now extinct, a presumption rendered all the more probable from the extensive deforestation which has occurred on that island; for the o-o is preeminently a bird of the dense forest.[48]

—H. W. Henshaw, 1902

A pair of Hawai'i 'Ō'ō, *Moho nobilis*, (above) dispute with a rival male (below) on a flowering 'Ōhi'a, *Metrosideros polymorpha* (Myrtaceae). (JULIAN P. HUME)

Hawai‘i ‘Ō‘ō

Moho nobilis

Description:

(from Gould) General plumage sooty-black; tail brown, all but the two middle feathers largely tipped with white; the two central feathers somewhat narrower than the others, and gradually diminishing in the apical third of their length into fine hair-like or filamentous upturned points; axillae or under surface of the shoulder white; flanks and under tail-coverts bright yellow; bill and legs black. The plumage of the female is in every respect similar to that of the male; but, as in the Honey-eaters of Australia generally, particularly amongst the members of the genus *Ptilotis*, the body is fully a fourth less in size.[49]

Measurements:

Total length 12 inches, bill 1½, wing 4¾, tail 6¾, tarsi 1⅕[50]

—Scott B. Wilson, 1899

Geographic range: Island of Hawai‘i.

Specimens in museums: 150.

Historical information: First specimen collected 1779. Last observation 1902.

"A specimen collected by M. L. Walton on 13 May 1902 (LACM 16983) near Ka'au Crater is the last substantiated report (Banko 1979, 1981); two uncatalogued mounted specimens at the Delaware Museum of Natural History appear also to have been collected by Walton on this date."[51]

Reports and observations:

...a bird with a long tail, whose colour is black, the vent and feathers under the wing (which is much longer than usually seen in the generality of birds, except for the birds of paradise), are yellow.[52]

—Captain Charles Clerke, 1779

The yellow tufts of costal feathers on this beautiful bird, furnished the material for the splendid and costly royal robes, capes, and "lei's," or head-dresses of the Hawaiians in former days. The bunches of feathers called *hulu,* ... are still prepared and received in payment of a poll tax to the King; they are afterwards made up primarily in "lei's," or headbands worn by the ladies, and are beautiful but costly ornaments; but few can afford to wear them. The mantles made up of these feathers, were, until lately, considered the principal treasures of the crown; now they are not to be seen; the labour of collecting the feathers and attaching them to a network base—a labour of years-being to great.[53]

—Titian R. Peale, 1848

This is, perhaps, the best known species both to natives and denizens of the Sandwich Islands; for it was principally from the yellow feathers that grow beneath its wings, together with the still more beautiful and similarly coloured upper tail-coverts of the now extinct *Drepanis pacifica*, that the state robes of the princes were fabricated. It was the privilege of those classes alone to wear them; nor can it be denied that they formed a becoming and magnificent garb, as beautiful as anything that the triumphs of civilized art can now produce.[54]

That large numbers of the O-O must have been taken in former days is clear from the quantities of "leis" or wreaths of feathers that now remain in the possession of the natives, who still set so great store by them that it is but rarely that a traveler is able to purchase so interesting a relic of a past age; but I was fortunate enough to obtain a small example, for the construction of which it is reckoned that two hundred birds must have been sacrificed.[55]

This bird is preeminently a honey-sucker, extracting the nectar with its long tubular tongue from the flowers of the ohia or from those of the great tree-lobelia, the hollow curving corolla of which is perfectly adapted in shape to the O-O's bill; and though I have on several occasions observed it feeding on the fruit of the banana, I believe this to be only a secondary article of diet: in a state of captivity it has been kept with success on the juice of the sugar-cane.[56]

Hawai'i 'Ō'ō, *Moho nobilis*. (SMITHSONIAN INSTITUTION)

F. W. Frohawk del. et lith. West, Newman imp.

ACRULOCERCUS NOBILIS

We shot two O-Os to-day, but these birds are extremely difficult to obtain, as they are constantly on the move from tree to tree, hardly ever at a less height than 90 feet from the ground. Their cry is somewhat harsh, and resembles the sound of the letter O repeated twice, with a well-marked pause between; it is, however, extremely difficult of imitation by the human voice. The yellow axillary tufts are very conspicuous when this bird is on the wing, and its dipping mode of flight somewhat resembles that of the Magpie, while its long tail still further suggests a resemblance to that bird. The O-O exhibits a decided preference for the extreme top of any tree, on which it alights, and when thus perched may be seen continually jerking its long plume-like tail up and down at a right angle to its body, all the while uttering its harsh cry." As mentioned above, it is an extremely wary bird and most difficult of approach when met with in the ohia-forest, so that the only occasions on which I was enabled to watch it at close quarters were amongst the foliage of the lobelias.[57]

The O-O is esteemed a great delicacy by the natives, and used formerly to be eaten by them, fried in its own fat. I can vouch personally for its excellence, as one day, after bringing in a fine specimen from a collecting-expedition, I placed it carefully on a shelf to await my convenience; but at dinner the Chinese cook, Ah Lung, set down in front of me a small dish, containing my lovely prize! On remonstrances being addressed to him in no measured terms, he only smiled and said "Me thinkee all same Kolea (Plover)." However my host, Mr. Spencer, and I tasted it, and found it excellent.[58]

—Scott B. Wilson, 1899

Though island royalty is no more, the demand for feather leis continue, and to meet it the o-o has been pursued till only a few remain in the distant woodland fastnesses-so few and so shy from persecution that happily the yield to the lei hunter no longer repays the labor and trouble of pursuit.[59]

As late as 1898 more than one-thousand individuals of this species were shot by the lei hunters in the heavily wooded district north of the Wailuku river, where their presence had probably been overlooked.[60]

J. G. Keulemans del. et. lith Mintern Bros imp.

MOHO NOBILIS, (MERREM)

When feeding in the early morning, and particularly when with their young, the calls of the o-o are almost incessant, and it is this loud and constantly repeated call-note which has led to the easy destruction of the species. The poor bird has yet to learn that its appreciation of the joyousness of existence and its love for its mate and young can be expressed only at the cost of its very life.[61]

—H. W. Henshaw, 1902

F. W. Frohawk del. et lith. West, Newman imp.

ACRULOCERCUS BISHOPI

Molokaʻi ʻŌʻō (Bishop's ʻŌʻō)

Moho bishopi

Description:

Adult male. Head and occiput deep black with a very slight metallic gloss, the shafts of the feathers a little paler. Neck, back, breast, and abdomen smoky black, with narrow white shaft-lines to the feathers. Hump and upper tail-coverts black. Wings and wing-coverts black. Tail black, with very narrow white fringes to the tip. Ear-coverts with an elongated tuft of golden-yellow feathers, these feathers black at lowest base. Axillary tufts bright yellow, shorter than in Moho nobilis. Under tail-coverts bright yellow. Iris dark brown. Bill and feet black. Soles dark flesh-colour with a yellow tinge.[62]

Measurements:

Total length about 11 to 11.75 inches, wing 4.5 to 4.7, tail 6.3 to 6.6, lateral tail-feathers 3.5 inches shorter, tarsus 1 to 1.5, culmen 1.4 to 1.45.

Adult female. Similar to the male in colour, but smaller. Wing 4 to 4.15 inches, tail 5 to 5.4, culmen 1.2, tarsus 1.45.[63]

—Walter Rothschild, 1900

Geographic range: Island of Molokaʻi.

Specimens in museums: 31.

Historical information: First specimen collected 1892. Last observation 1904.

Reports and observations:

On December 16th Palmer was guided by Dr. Mouritz, who showed much kindness to him in many ways, to one of the highest peaks on the island and there it was where he shot the first specimen of that beautiful new "O-o" which I named *Moho bishopi.*[64]

The first week of January 1893 was almost entirely spent in search of the *Moho bishopi*, and as many as five specimens, besides some other birds, were shot, which Palmer considers a very good success…[65]

J. G. Keulemans del. et. lith Mintern Bros imp.
MOHO BISHOPI, (ROTHSCH.)

Palmer writes: "This bird is rather shy when approached by men, but sometimes it is rather inquisitive, if one keeps quiet, and it generally answers and comes up to the spot when its cry is imitated.[66]

—Walter Rothschild, 1900

The Moloka'i or Bishop's Oo *(A. bishopi)* does not greatly differ in its habits from the Hawaiian bird, except that in its case these habits are probably more natural, either because it was less hunted by the natives, or because it ceased to be persecuted long before the Hawai'i species, owing to its diminished numbers. That it was hunted to some extent is certain from the fact that in the possession of an old native woman I found a large collection of the yellow feathers of this bird, some loose and some already tied in bunches with Olona fibre. That these had been obtained from birds caught alive and not shot was evident, as many feathers still showed traces of the birdlime. Bishop's Oo is a great nectar-eater, and it visits the *Lobeliaceae*, or rather certain species of these, in preference to any other flowers. At times it may be seen numerously in the Ohia trees, when these are in flower, but this is clearly because its favourite lobelia nectar is unattainable. Thus on one of my visits to Moloka'i of about a score of the Oo that I watched while they were feeding, at a time when the Ohias and lobelias were both in flower, one only was seen to visit the former, though these were attractive enough to other birds. This Oo also visits the flowers of the banana in some of the deep gulches and valleys, and is said to also eat the fruit. To insect food, when the nectar of the lobelias is abundant, it pays little attention.[67]

When seen in the low brush of the Molokai forest the Oo is a somewhat inquisitive bird, but restless and timid when nearly approached, far more so than are most of the other forest birds that are found with it. It is easily called by imitating its cry, though it will not infrequently come and inspect the intruder uncalled. At such times the mixed timidity and curiosity of its nature are nearly always shown, and it rarely exposes itself to more than momentary view, diving beneath the foliage of the bushes at the faintest alarm, and sometimes only thrusting out its head and neck from beneath the leaves to renew its inspection. The cry of the Oo is unlike that of any other native bird, and no one who has once heard and identified it can ever again be in doubt as to the bird that utters it. This cry is usually dissyllabic, as represented by the bird's name, but sometimes may be represented by only a single sound. It varies somewhat, especially in loudness and clearness, according to the season, and is either uttered once or repeated. Thus at certain seasons the usual cry of the Molokai bird at its fullest was well represented by five syllables, owow, owow, ow, the first and the second, as likewise the third and fourth, are practically continuous, with the stress on the latter in each case; there is a slight pause between the first and second pair of syllables and generally a longer and distinct one between the latter of these and the final one, which is extremely loud and might at times be almost termed a shriek. These cries are naturally audible at a great distance, and I have estimated that I have heard the cry of the Hawaiian species when half a mile from the bird, and that of *A. bishopi* is certainly sometimes audible at a distance of at least 3000 feet. When the former assemble in large numbers in the flowering Ohias, their cries are incessant, and are uttered not only from the trees but sometimes even on the wing, as the birds continually pass backwards and forwards from one tree to another. In fine weather they are mostly silent during the hotter hours, but in wet or foggy weather are often vociferous throughout the day.[68]

—R .C. L. Perkins, 1903

Endemic to Molokai and in danger of extinction, Bishop's oo was discovered by Palmer in 1892 at Kaluaaha, Molokai. Rothschild named it after Mr. Charles R. Bishop who founded the Bishop Museum. It frequented the upper forest of Molokai. Perkins was able to make a good study of it. He found that the natives used it for its yellow plumes. An old woman showed him a number of these feathers, some loose and some tied in bundles with olona fiber. He traced marks of the bird lime with which the birds were snared, on the feathers. Members of the Meyer family collected a number of birds in the 1880's and have some still in their possession. I saw a group of about half a dozen in 1904, the last authenticated instance of its being seen.[69]

—George C. Munro, 1944

Professional bird catchers serving for King Kamehameha or local chiefs were under strict license to provide feathers for the spectacular feather capes and helmets. Teams of up to twenty-five men, accompanied by their wives and wearing rain capes made from ti leaves, spent weeks in the mountain forests.[70] To catch the birds, birdlime, a sticky glue made from tree sap, was spread onto a branch with a decoy ʻIʻiwi *Drepanis coccinea* tied alongside.[71] The yellow plumes of ʻŌʻō *Moho* sp. were much sought after, and as these birds reacted aggressively towards ʻIʻiwi, they became mired by the lime. Possibly as a sign of status, the bird catchers never cut their toe and fingernails, which grew like claws.[72] (JULIAN P. HUME)

Previous page: Three species of undescribed fossil Mohoidae squabble above the Puʻuʻulaʻula Crater, Maui. Maui ʻŌʻō *Moho* sp. (left) is driven away from a flowering ʻōhiʻa by a narrow-billed Kioea *Chaetoptila* sp., while a Kioea *Chaetoptila angustipluma*, or something very closely related to it, looks on. (JULIAN P. HUME)

Maui ʻŌʻō

Moho sp.

Description: None from specimens.

Measurements: None from specimens.

Geographic Range: Island of Maui.

Specimens in museums: None.

Historical information: First observation 1901.
Last possible observation 1986.

Reports and observations:

"The evening before we left, our host proposed to take us home over the mountains, instead of the canoe route, to which we assented, as offering a little adventure. A company of twenty-five athletic men, trained to bird-catching on the beetling crags of these mountains, were called out. Their toe and finger nails, never cut, grow like claws. Their sole business is to catch the little black birds called the *oo,* each producing a few yellow feathers under the wings. When these are plucked, the bird is set at liberty, to be caught and plundered again at some future time. Five feathers constitute a tax, and are equivalent to one dollar in money. These feathers are wrought into cloaks, capes, wreaths, and "kahilis" to bedeck royalty. The feathers, time, and labor consumed in making a cloak, now in possession of the young king, have been estimated as amounting to hundreds of thousands of dollars.[73]

—Laura F. Judd, July 1828

"The species may yet, however, be found to occur on the island of Maui from which, hitherto, no species of *moho* has been reported. On June 9, 1901, however, I saw a single adult male O-O in the forest tract northeast of Olinda, at an elevation of about 4,500 feet. I was close to the bird, saw it plainly, and heard its notes distinctly, and am positive as to its identification as one of the members of this genus. The species has yet to be determined but, if not an undescribed species, it is likely to prove to be *bishopi*, and to be common to both Maui and Molokai, as is the *Palmeria dolei*."[74]

—H. W. Henshaw, June 1901

"I was on a small ridge at 1900 m elevation in the Ko'olau Forest Reserve on 10 May 1981, when at 10:00 HST I heard a series of unfamiliar bird songs about 150-250 m away. These songs each comprised 2-3 repetitions of a pair of spondee-accented notes "owh - owh" with low pitch and soft, rich tonal quality. About 10 such songs were given over a 5 min interval. I stalked towards the songs, and came to a sheer gulch overlook at 1850 m elevation. Here I surveyed the 'ohi'a-lehua (*'Metrosideros collina*) forest along the gulch walls for 3 hours. At 11:15 I saw a long-tailed black bird fly into the subcanopy of an 'ohi'a tree below me from the understory of *Cheirodendron trigynum*, *Coprosma ochracea*, and *Styphelia tameiameiae*. The bird moved actively from branch to branch in short flights, perched upright for about 10 seconds, and then flew away at subcanopy level. Dense understory and steep topography prevented pursuit. I had excellent dorsal, lateral, and frontal views of the bird through 8x30 binoculars during about 30 seconds of observation from a distance of 30-40 m. The weather was clear and the sun was behind me. At 11:45 I heard 5 songs, identical to those heard earlier, given in the gulch. These songs and sighting were the only ones of their kind encountered during 10 continuous field days in the vicinity. The following description is taken from on-the-spot notes and sketches: "Total length 25-30 cm. Bill about 4 cm long, black, slightly decurved, not compressed. Eye dark. Body and wings smoky black, crown with gloss; ear coverts cadmium yellow, tufted appearance; axillary wing tufts 2-3 cm long, cadmium yellow; crissum appeared yellow but not seen clearly. Legs 5-6 cm long, black. Tail black, much graduated, very elongate central feathers about 15 cm long, streaming in flight". One other medium-sized all-dark bird, the immature Crested Honeycreeper (*Palmeria dolei*), also occurs in these forests. It differs from the 'O'o by its smaller size, stouter appearance, normally proportioned tail and lack of yellow patches."[75]

—Stephen R. Sabo, 1981

"Moho sp.? presumably endemic to Maui, is known solely from a single field observation in 1901. No convincing re-sighting of this genus has been reported from Maui in the past 79 years. Specific characteristics of the form seen and heard were not mentioned in the type observation and are wholly unknown. Exhaustive search of literature and field journals for information on population status and geographical distribution uncovered some 17 reports and related statements dated from 1828 to 1980. Records are serially numbered, arranged in geographical and chronological order, referenced, and systematically reviewed. The positive identification of an 'O'O in 1901 is recognized. There is some evidence that an undescribed species of *Moho* (or *Drepanis*) existed on Maui at least until 1980."[76]

—Winston E. Banko, 1981

"'O'o have been reported from Maui since 1828 (Banko 1980-1984), the most notable sightings being by Henshaw (1902) and Sabo (1982). In these two sightings the observers were convinced that the bird was an adult 'O'o , possibly Bishop's 'O'o from the field marks; these records have been accepted by Pyle and Ralph (1982) and the A.O.U. (1983). Until a specimen or photograph is obtained, however, the specific identity of the "Maui 'O'o" remains debatable. The most recent sightings were from ohia rainforests on the northeast slopes of Haleakala, in the Hanawi watershed at 1600-2000 m elevation. There was one incidental sighting of an unidentified black bird with an 'O'o silhouette from that area during the 1980 HFBS and another by D. Boynton (pers. comm.) in 1983. Fossils of *Moho* sp. occur on Maui (S. L. Olson, pers. comm.)."[77]

—J. M. Scott et al., 1986

"The possible occurrence of Bishop's 'O'o or a similar congener on Maui has generated much discussion and interest (Banko 1981, Sykes 2000) but no substantiated records. There is subfossil evidence for a *Moho* on Maui (James and Olson 1991) and Bishop's 'O'o certainly could have inhabited Maui Nui when Moloka'i and Maui were connected (Wilson and Evans 1899, Sabo 1982). Judd (1880) accompanied bird-catchers to the W Maui Mts on 4 Jul 1828 in search of "the little black bird called the oo, each producing a few yellow feathers under the wings". Sight and aural observations of black birds have been reported from the Waikamoi and Hanawi watersheds on the e. slopes of Haleakala, in June 1901 (Henshaw 1902); Nov 1950, 1955, Sep 1973, and May 1974 (Banko 1981, E 21:26); May 1980 (Sabo 1982); and several times in 1981-1986 (Scott et al. 1986, Sykes et al. 2000). Extensive ornithological surveys in these areas from 1986-2005 has failed to turn up any further evidence of this conspicuous genus."[78]

—R. L. Pyle, and P. Pyle, 2017

The ʻŌʻō and the Mamo are birds that have a great resemblance to each other. They are smaller than the ʻŌʻū, have black feathers, sharp beaks, and are used for food. Their feathers are made into large royal kahilis. Those in the axillae and about the tail are very choice, of a golden color, and are used in making the feather cloaks called ʻahuʻula which are worn by (the aliis as well as by) warriors as insignia in time of battle (and on state occasions of ceremony or display—Translator.) They were also used in the making of leis (necklaces and wreaths) for the adornment of the female chiefs and women of rank, and for the decoration of the makahiki idol. These birds have many uses and they are captured by means of bird-lime and the pole."[79]

—David Malo, 1839

The forty-nine Hawaiian feather cloaks and capes above are only 28 percent of the 171 that exist. The number of birds required to construct a single cloak range from 7,000 to 20,000. Red feathers were obtained from the ʻIʻiwi and ʻApapane, green from the ʻŌʻū and yellow from the ʻŌʻō and Mamo.[80]

F. W. Frohawk del. et lith. West, Newman imp.

CHÆTOPTILA ANGUSTIPLUMA

Kioea

Chaetoptila angustipluma

Description:

Top of the head and neck blackish brown, each feather with a grayish white shaft-streak, which is strongly tinged with yellow on the nape and sides of the neck. A greyish-white stripe over the eyes. Wing-coverts and back hair-brown, tinged with ochreous on the rump, the feathers of the mantle with a white shaft-streak widening into a tear-shaped spot towards the tip. Remiges and rectrices deep brown, their outer margin yellow, giving a greenish effect to the whole. A greyish-white stripe over the eye. Lores, sides of the head, and ear-coverts dull black, the feathers immediately under the eye mottled with greyish white. Chin and throat dull white, tinged with yellow, the shafts and the hairs with which this part is beset black. Breast and abdomen dull white, striped longitudinally with darkish brown, flanks strongly tinged with ochreous. Bill and legs very dark brown, almost black.[81]

Dimensions:

Total length about 13 inches, wing 5.5, tail 6, bill (allowing for a slight injury at the tip) from forehead 1.31, from gape 1.5, tarsus 1.75, middle toe without claw .93, hind toe .56.[82]

—Scott B. Wilson, 1899

Geographic range: Island of Hawai‘i. The same or similar species occurred on O‘ahu and Maui.

Specimens in museums: 4.

Historical information: First specimen collected 1840. Last observation 1859.

Reports and observation:

This rare species was obtained at the Island of Hawaii. It is very active, and graceful in its motions, frequents the woody districts, and is disposed to be musical, having most of the habits of the *Meliphaga*; they are generally found about those trees which are in flower.[83]

—Titian R. Peale, 1848

J. G. Keulemans del. et. lith Mintern Bros imp.
CHÆTOPTILA ANGUSTIPLUMA, (PEALE)

The feathers on the head and breast in this bird present a remarkable character on account of the filaments composing the webs of the feathers being unusually few in number, and at such a distance from each other as not to touch, or become adherent. This structure of the feathers gives to the plumage of the parts mentioned a somewhat hairy appearance, and prevails also, in some measure, on the abdomen and other underparts of the body. Many of the feathers on the throat and neck in front terminate in bristles curved outwardly, and readily discernible on viewing the specimen in profile.[84]

Dr. Pickering mentions having seen this species "alighting in the tops of the trees and uttering a loud *chuck*."[85]

—John Cassin, 1858

I obtained an example of this curious-looking bird from the collection of the late Mr. Mills, through the generosity of the present owner, the Hon. C. R. Bishop, of Honolulu, and it is now in the Museum of the University of Cambridge. It was probably procured by Mr. Mills in the district of Hilo near Olaa, that having been his favourite hunting-ground. Why this bird should have become extinct seems inexplicable, as its feathers were not used for ornaments; yet the natives of the present day do not know it even by tradition, moreover the local name Kiowea given to it by Judge Dole is certainly that applied to the Whimbrel *(Numenius femoralis)*.[86]

—Scott B. Wilson, 1899

Chaetoptila Angustipluma is evidently extinct. Nobody in the Sandwich Islands seems to know anything about it, and its former native name is unknown. Besides the specimen in my collection, we only know the type of the species in Washington, one in the Cambridge Museum, from the same source as my example, and one in the Honolulu Museum. All these specimens came from Hawaii. Dole quotes also Molokai as the habitat of this bird, but we do not know why. All these specimens (except the type) were procured by Mills, who, according to Palmer's enquiries, collected them between the lower Volcano House and the crater of Kilauea.[87]

—Walter Rothschild, 1900

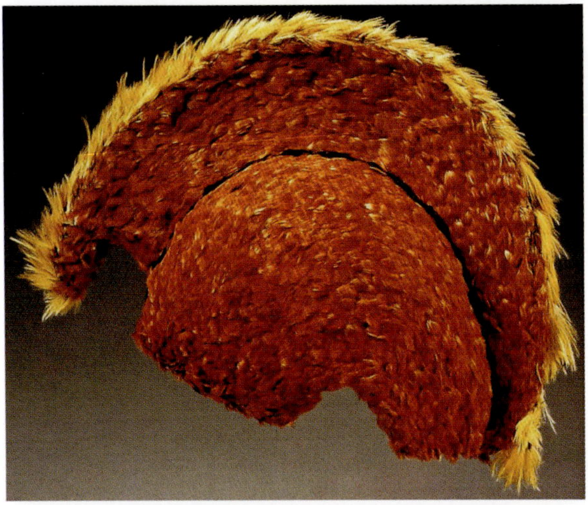

Mahiole—A Hawaiian feathered helmet that was given to Captain James Cook by Kalani'opu'u a high chief on the island of Hawai'i in 1779. The red feathers are from the 'I'iwi *(Drepanis coccinea)* and the yellow feathers are from the 'Ō'ō *(Moho nobilis)*. (COURTESY OF THE MUSEUM OF NEW ZEALAND TE PAPA TONGAREWA)

Scaled illustration of all the historically known Mohoidae. The entire family is now extinct. Clockwise from top: Moloka'i 'Ō'ō, *Moho bishopi*; Kioea, *Chaetoptila angustipluma*; O'ahu 'Ō'ō, *Moho apicalis*; Kaua'i 'Ō'ō, *Moho braccatus*; Hawai'i 'Ō'ō, *Moho nobilis*. (JULIAN P. HUME)

Laysan Millerbird

Acrocephalus familiaris familiaris

Description:

Adult male. Upper parts grayish brown, rather warmer in tinge on the rump, the sides of the head rather paler; wings and tail dark brown, the feathers externally margined with buffy brown; tail rounded; wings somewhat rounded, the first primary shorter than the sixth. Underparts, together with the chin and throat, buffy white. Bill brown; legs fleshy brown; tarsus brown.
Adult female- Closely resembles the male.[88]

Measurements:

Total length about 5.75 inches, culmen 0.65, wing 2.55, tail 2.4, tarsus 0.92.[89]

—Walter Rothschild, 1900

Geographic range: Island of Laysan.

Specimens in museums: 78.

Historical information: First specimen collected 1891.
Last observation 1923.

Reports and observations:

The "Miller-bird" is plentiful on Laysan and very tame. All specimens sent were caught with a hand-net. It is very active in its habits, hopping and creeping among the grass and scrub in search of insects. It takes its name from its fondness for some large white moths, called on the island "Millers." For the size of the bird it has a harsh deep note, much resembling that of a Thrush. It lays three eggs, and the nest is placed in a tussock of grass.[90]

—Walter Rothschild, 1900

The Miller Bird is one of the most abundant of the species under consideration and is seen to best advantage during the cool of the morning or in late afternoon, for then it is very active, and at times musical. During the heated portion of the day, after the custom of our wood warblers, it retires to remain in seclusion among shady bushes, or tall tussocks of grass where its nest is made. Like most of the birds on the island, *Acrocephalus* is rather unsuspicious though not by any means so tame as either the Finches or Rails.[91]

—Walter K. Fisher, 1893

J. G. Keulemans del. et. lith Mintern Bros imp.

ACROPEPHALUS FAMILIARIS, ROTHSCH.

With the exception of the Laysan teal, the miller bird is the least abundant of the indigenous birds. We saw a few of them around the old buildings and others about the island, but principally along the west shore of the lagoon, in tall grass, where a few nests containing eggs and young birds were found. It seems strange that this bird should not be more common, as it was formerly reported to be most abundant of the land birds.[92]

—Homer R. Dill, 1911

My surprise was great, however, to find that the few rabbits introduced by Mr. Schlemmer shortly after my former visit had literally taken possession of the island. In 1903 Mr. Schlemmer, then operating a lease engaged in the guano business, made a number of trips from Honolulu. On a number of these voyages he took several small shipments of rabbits of different breeds as well as a number of Japanese guinea pigs. All of these animals were liberated on Laysan and they have since multiplied until they are found in great numbers at this time.[93]

All of the small species peculiar to the island, except possibly the Laysan teal, depend entirely upon the vegetation on the island for their food supply. The uninterrupted and astonishing increase in the numbers of rabbits and guinea pigs can have but one result if allowed to continue unchecked. They will surely eat or kill off all of the grass and shrubs. Doubtless the insects that feed on the plants and in turn are fed upon by the birds will be so reduced as to bring about starvation among the small birds.[94]

The Laysan canary, Miller bird, Laysan honey eater, and the Laysan Crake are doomed to extermination on the isolated island on which they have maintained themselves long enough to develop into distinct species, unless something is done to preserve for them the source of their food supply.[95]

—William A. Bryan, 1911

Grave fears for the smaller birds passed through my mind and it seemed inevitable that the teal, millerbird, rail, *Himatione* and finch had perished.[96]

But not a trace of rail or honeyeater or millerbird could we find.[97]

—Alexander Wetmore, 1923

‘Amaui (O‘ahu Thrush)

Myadestes woahensis

Description:

Upper parts olive-brown, extremities of the feathers much lighter color; tail and wings brown; bill bristled at the base.[98]

Measurements:

Length 7½ inch.[99]

—Andrew J. Bloxam, 1825

Geographic range: Island of O‘ahu.

Specimens in museums: 2.

Historical information: First specimen collected May 14, 1825. Last observation May 14, 1825.

Reports and observations:

We soon began to ascend the pass, the sun rising at the time, amid the chirpings of small birds and the melodious notes of a brown thrush, the only songster on the islands.[100]

—Andrew J. Bloxam, 1825

Thanks to the diligent inquiry of Mr. Wilson we now know that the island of Oahu formerly possessed a member of this family of thrushes, and that in 1824 specimens of the bird were obtained on the island by Bloxam, the naturalist of the "Blonde," who gave it the manuscript name of *"Turdus woahensis"* (see *Wilson Birds of the Hawaiian Islands* p. XIII.)

The bird doubtless became extinct long ago, and it seems never to have been seen, or at least recorded, from Bloxam's day to this.[101]

As Bloxam's manuscript contains a description of the thrush inhabiting the island of Hawaii *(P. obscura)* which he rightly considered a different species, we are safe in believing that the Oahu bird was fully entitled to a distinct name, although no specimens of the bird are now known to be in existence.[102]

—Henry W. Henshaw, 1902

The extinction of the Oahuan species of the genus is a most singular fact, when one considers that *P. Lanaiensis* was found so commonly in the much smaller forest area of Lanai, and on Molokai still remains in spots whence the other native birds have vanished. It may be, however, that the Oahuan bird had very special habits, as appears to be the case with the small Kauai species, and that for this reason its range was also very limited. The fact that it does not seem to have been observed by any of the old collectors except Bloxam would lead one to suspect this to be the case.[103]

—R. C. L. Perkins, 1903

Skin study of the probable type of O'ahu Thrush, *Myadestes woahensis*. Specimen is in Tring, UK. (JULIAN P. HUME)

F. W. Frohawk del. et lith. West, Newman imp.

PHÆORNIS MYIADESTINA
PHÆORNIS LANAIIENSIS

Lānaʻi Olomaʻo (Lānaʻi Thrush)

Myadestes lanaiensis lanaiensis

Description:

Adult male. Above dusky olive-brown, the head more blackish brown. The wings like those of *Ph. myiadestina*, Stejn., but the outer edges of the primaries more brown and less rusty, the bases of the same less rufous. Chin, throat, and breast pale grey, with a faint isabelline tint, passing into the nearly pure white abdomen. Under tail-coverts pale huffy white. Outer pair of rectrices brown; outer webs brownish buff with a very narrow outer marginal line and a small tip of pale cinereous, but without white. Middle rectrices the same colour as the back, the following ones edged with the same colour on the outer webs. Iris hazel; bill black; tarsi and feet dark brown, soles yellowish.[104]

This species is easily distinguished from *Ph. obscura* (Gm) by its much smaller size and whiter underparts, and from *Ph. myiadestina*, Stejn., by the absence of the conspicuous white markings in the tail and the smaller size. The bill is somewhat between those two species in width, but varies a little, and it is not so much pointed as in *Ph. obscura.*[105]

The *adult female* is not to be distinguished from the male if my birds are properly sexed. The young bird is like those of the allied species, and I trust Mr. Keuleman's well-executed figure will give a better idea of it then a long description.[106]

Measurements:

Total length about 7½ inches and more as measured by Palmer, but only about 7 in skins. In fourteen specimens from Lānaʻi the wings measure 3.49 to 3.75 inches (average 3.65), while in thirteen from Molokai they measure 3.65 to 3.9 (average 3.8), tail 2.9 to 3.1, tarsus 1.3 to 1.35, culmen 0.6, breadth of bill just behind nostrils 0.27 to 0.3, middle claw without toe 0.75.[107]

—Walter Rothschild, 1900

Geographic range: Island of Lānaʻi.

Specimens in museums: 44.

Historical information: First specimen collected 1888.
Last observation 1934.

Reports and observations:

I obtained several specimens of this bird on the island of Lanai, which from its much smaller size and the whiteness of the underparts is deserving of separation from either *Phœornis obscura* of Hawaii or *P. myiadestina* of Kauai. It appears to be identical with the species inhabiting Molokai,—as was to be expected, seeing that the two islands are separated only by a narrow channel, some ten miles in width—and is known by the name of Olomao there: on Lanai, where I first obtained it, and after which island it is named, I met with no natives who knew the names of birds; indeed, in a few years there will not be many natives remaining.[108]

—Scott B. Wilson, 1899

The specimens from Lanai, the island from which Wilson's type came, are as a rule much whiter below, and the majority of them have brown of the back somewhat less bright.[109]

—Walter Rothschild, 1900

It inhabited all the present forest, frequenting the low trees and underbrush and, unlike the other species, did not seek the highest dead trees to broadcast its few call notes which replaced the beautiful song of the other thrushes. From 1911 to 1923 this bird was under my observation as I frequently rode the bridle trails of the forest. It was at that time a common bird and its call notes could be heard constantly, especially in the north and south ends of the small Lanai forest. It declined from 1923 when the population of Lanai increased and the town of Lanai City was built. The people brought bird diseases with their poultry and these, evidently carried by mosquitoes, were fatal to the native bird population. I watched its decline till 1931. The few times I have been through the Lanai forest since 1931 the thrushes call notes have been conspicuously absent.[110]

Its food is berries and insects. I found a small land shell in one.[111]

—George C. Munro, 1944

J. G. Keulemans del. et. lith Mintern Bros imp.

(1) PHÆORNIS OBSCURA, (GM.) JUV.
(2) PHÆORNIS MYIADESTINA, STEJN. JUV.
(3) PHÆORNIS LĀNAʻIENSIS, WILS. JUV.

Moloka'i Oloma'o (Moloka'i Thrush)

Myadeastes lanaiensis rutha

Description:

Adult Male: Above inconspicuous dull brown with dusky olive tinge; head darker colored than the mantle. The grayish wash of Lanaiensis over the head and rump wanting even in moulting birds. Outer webs of the inner primaries and most of the secondaries with a blackish patch at their bases, which is bordered before with a rusty brown patch; centre tail feathers like the back; outer pair paler on their outer webs on the basal portion, but not conspicuously so. Lores blackish and gray, grayest on the breast and palest on the throat; sides of the body like the breast. Flanks and thighs rusty olive. Centre of the abdomen clear white; under tail coverts ranging from pale buffy cream to white; bill black, iris brown; feet and tarsus dark brown; soles pale yellow in life.

Adult Female: Similar to the male in color. Immature Male: Similar to young Lanaiensis, but much darker and less rusty olive over the back; under parts uniformly grayer; wings and tail longer.[112]

Dimensions:

Length 8.25 (8.31)-8.40, wing 3.67 (3.73)-3.80, tail 3.10 (3.17)-3.25, tarsus 1.22, toe 1.00, culmen 70.[113]

—William A. Bryan, 1908

Geographic range: Islands of Moloka'i and Maui.

Specimens in museums: 52.

Historical information: First specimen collected 1888.
Last observation 1980.

Reports and observations:

For the Olomao is by nature a shy, timid bird, and for that reason, instead of exposing itself in the open it prefers to proceed from place to place by short low flights, usually through the shade of the forest.[114]

Berries are by far its commonest food. While insects were found in the stomachs of a part of the birds examined, they occurred in no considerable quantity in any of them.[115]

A bird under close observation flew down to feed on the drupes of a small Olapa (*Cheirodendron*). After a few moments it flew up into a nearby tree, when after deliberately cleaning its bill it broke forth into its fullest song. Pausing as if to study the effect, or to see if the melody would be taken up by its neighbors, it would tremble the wings and hop idly about from branch to branch. This program was repeated many times, singing, feeding, and fluttering its wings alternatively.[116]

The effort is more usually delivered from the topmost branches of some favorite tree; although it is to be found frequently singing joyously in the underbrush. When singing the head is always thrown well back, the throat full and free, and the wings and tail are invariably relaxed and drooping. Once heard its character will live in the memory for years, though its component parts are wholly inimitable.[117]

—William A. Bryan, 1908

Endemic to Moloka‘i, it was common in the eighteen-nineties, but is now in danger of extinction. Perkins found it still inhabiting the outer forest after the other native birds had deserted the locality. Bryan collected specimens in 1907. I saw it that year but it did not seem numerous as formerly. In the bird survey in 1936 I thought I heard one sing but was not absolutely sure. Donaghho went over a great deal of Moloka‘i forest in 1937 but did not see it.[118]

—George C. Munro, 1944

The last substantiated observations were of two near Pu'u Waha'ulu in Jul 1975 (Scott et al. 1977), 2 there in summer 1979 (Kepler and Scott 1980), and 3 on Olokui Plateau (between Wailau and Pelekunu valleys) in summer 1980, during the HFBS (Scott et al. 1986); we consider reports from 1988, 1994, and 2005 (USFWS 2006, 2009; BLI 2016) as unsubstantiated. Based on HFBS data, a population of 19 individuals was estimated in 1980 (Scott et al. 1986). We presume this population is extinct, although no fieldwork has occurred on near-inaccessible Olokui Plateau since the last observations were made there (Reynolds and Snetsinger 2001) and some still consider it "possibly extinct" (e.g., Scott et al. 2008, USFWS 2009, BLI 2016).[119]

—R. L. Pyle and P. Pyle, 2017

F. W. Frohawk del. et lith.

West, Newman imp.

PHÆORNIS MYIADESTINA
PHÆORNIS LANAIIENSIS

Kāma'o (Large Kaua'i Thrush)

Myadestes myadestinus

Description:

Adult male. Above dull hair-brown with an olive tinge, different from the color of the back of *Phaeornis obscura*, and somewhat more olive; sides of the head dull tawny, bordered with dusky. Quills deep brown, edged externally with rusty olive, but the second to the fifth greyish brown near the tips. The inner primaries and most of the secondaries have a very peculiar pattern; the shaft is brown above, whitish below; the inner webs are dark brown, except on the basal third, which is very pale buff. The outer web is bright russet, except near the tip, where it is only margined with russet, and crossed behind the middle by a blackish band. Lower parts of a light smoky grey, lighter than in *Ph. obscura,* and somewhat motley in appearance, of which character nothing is seen in *Ph. obscura*, shading into white on the vent. Middle tail-feathers like the back three outer tail-feathers with a white mark at the tip, very long and gradually shading into the brown on the outer pair, very small on the third pair. Under tail-coverts pure white. Bill black; iris brown; tarsi and feet dark brown, soles pale yellow.[120]

The adult female is entirely similar to the male in colour, and my series does not bear out any constant difference in size.[121]

Young birds are similar in colour to those of *Phaeornis obscura* (Gni.), but are distinguished from them by the whitish marks on the outer tail-feathers and the much broader bill, the latter characterizing the Kauai bird.[122]

Measurements:

Total length nearly 8 inches, wing 4.05 to 4.25, tail 3.3 to 3.6, culmen 0.57 to 0.6, breadth of bill just behind nostril 0.35, middle toe without claw 0.7, tarsus 1.2 to 1.25.[123]

—Walter Rothschild, 1900

Geographic range: Island of Kaua'i.

Specimens in museums: 47.

Historical information: First specimen collected 1835.
Last observed February, 1989.

Reports and observations:

This species, which inhabits Kauai, has been lately described as new by Dr. Stejneger. I obtained specimens on that island from nearly the same locality as he did, several being in immature plumage. Dr. Stejneger gives as native names Ou or Uapauau, neither of which is, as far as I know, applied to this bird, Kamao being that by which it is generally known.[124]

—Scott B. Wilson, 1888

Once more among the forest and its feathered tribe! What a striking contrast from Oahu! Everywhere here the trees appear to be alive with birds. The noise of their various songs was quite deafening this afternoon. The 'boom' of the 'O-o' *(Moho braccata)* and the powerful song of the 'Kamao' *(Phaeornis myiadestina)* were very noticeable above the others.[125]

—Henry C. Palmer, June 16,1893

The Kauai Thrush is not rare on Kauai. According to Mr. Knudsen it sometimes sings on the wing, as before stated of *Phaeornis obscura* (Gm.). Palmer describes it as being a quiet bird and not shy. Its rather melancholy and often repeated call-note attracts one's attention. Its song reminded him of that of an English Thrush, but it was less powerful, although it could be heard at a great distance, and, in Palmer's opinion, was sweeter. Its food consists of berries, but caterpillars and insect-remains were also found in its stomach.[126]

—Walter Rothschild, 1900

While the birds are shy, their size together with their song, which is given at all times of day and on one occasion was heard in the dead of the night-make them not a difficult bird to secure.[127]

—William A. Bryan and Alvin Seale, April 1900

It was extremely common in 1891 over all the forest region of the island from near sea level on the north side and outer edges of the forest to the mountain tops. In fact it was the most common bird in the Kauai forest at that time. It was still numerous on the forest edges in 1899 when the writer left Kauai.[128]

I visited the Kauai forest in September 1928, April 1931, July and August 1932, when I searched all the outside forests in vain for this bird.[129]

...In January 1936, in a visit to Kaholuamanu at about 3,700 feet elevation, a region I had not visited since 1899, I found this bird fairly common, with others of the Kauai birds.[130]

On a dead calm day I noted that the Kamao would sit with wings quivering within a few feet of me, turning its head on one side with a look of mild curiosity.[131]

—George C. Munro, 1944

We were sure that we had seen or heard more the 28 Kamao in a round trip over some four miles of trail, we feel there must be at least some hundreds, if not a few thousands, of these thrushes living in the Alakai swamp native forest area.[132]

—Frank Richardson and John Bowles, August 16, 1960

The state Forestry & Wildlife survey of three areas in the Alaka'i wilderness Feb. 12-17 (1989) had excellent weather but disappointing results. Only 3 observations (2 definite, one tentative) of the Kama'o were recorded, all in one area where many more had been recorded on the previous survey.[133]

—Robert L. Pyle, 1989

"We consider subsequent reports through 1995 (cf. USFWS 2006; Hawaii Wildlife Newsletter(HFW) 4[2]:10) as unsubstantiated. Extensive searches for endangered species in the Alaka'i Swamp in the 1990s failed to turn up any Kama'o (E 53:49; HFW 7[3]:1, 8[1]:12; Wakelee and Fancy 1999, Reynolds et al. 1997b; Reynolds and Snetsinger 2001, Foster et al. 2004, USFWS 2006, Gorresen et al. 2009) and we presume that the Kama'o is now extinct (see also Scott et al. 2008, BLI 2016). Based on Poisson analyses of persistence probabilities using confirmed and unconfirmed records, Elphick et al. (2009) estimated that the Kama'o went extinct in 1991-1999, with upper limits of 1999-2030."[134]

—R. L. Pyle and P. Pyle, 2017

F. W. Frohawk del. et lith. West, Newman imp.

HIMATIONE MONTANA

Lānaʻi ʻAlauahio

Paroreomyza montana montana

Description:

Adult male. Forehead, sides of the face, and throat deep lemon-yellow, shading into a lighter tint of yellow on the breast and abdomen, the lower part of which is white; under tail-coverts deep lemon-yellow; upper parts, with the exception of the rump, which is yellow, are dull greenish yellow; primaries (of which the second is much shorter than the fourth and fifth, which are equal) ashy brown, edged with dull yellow; wing-lining white, tinged with clear yellow; tail-quills ashy brown edged with dull yellow; bill light pinkish; feet slender, of the same colour as the bill.

Adult female. Similar in general colour to the male, though the underparts are of a very light shade of lemon-yellow instead of the deep yellow of the male.

Dimensions:

Total length 4 inches, wing from carpal joint 2.25, culmen .35, tarsus .70, tail 2.75.[135]

—Scott B. Wilson, 1899

Geographic range: Island of Lānaʻi.

Specimens in museums: 65.

Historical information: First specimen collected 1888.
Last observation 1937.

Reports and observations:

This hitherto undescribed species I obtained in the mountain-region of Lanai, at a spot called Lanaihale (the house of Lanai), at a height of about 3000 feet, the brilliant yellow of the underparts in the freshly killed male and its nearly straight bill clearly showing it to be distinct from any other member of the genus. I only met with four specimens, of which I secured an adult male and female on the same day in the locality mentioned above, and subsequently two immature examples in a gulch at a much lower elevation. As, perhaps, an account of our mountain trip on the day on which I shot the former may be of interest, I here transcribe from my Journal some notes, taken on the spot:—

1st June, 1888.—... I caught sight of a bright yellow bird in an ohia bush, a few yards down the side of the gulch; I put my gun instantly to my shoulder and fired, and down came the bird; F. and I scrambled down the gulch and fortunately found it. Its breast was of a brilliant yellow, far brighter in tint than the plumage of any other species I had previously obtained; its legs and bill were a light pink: in dissecting it I found some small larvae.[136]

—Scott B. Wilson, 1899

Palmer found this species on the south-west side of Lanai at elevations of about 1500 feet to the uppermost peaks. The call-note is a short and plain "tweet.... It is perhaps not so very rare, for Palmer sent me a fine series, but the dense bush and broken country make it very difficult to obtain.[137]

—Walter Rothschild, 1900

Endemic to Lanai, and almost certain of complete extinction.[138]

It was formerly common in all parts of the Lanai forest, but has declined in late years with the other Lanai birds. A pair was seen in March 1937 but it had been scarce for some time before that. There seems little or no chance of its survival and it will probably share the fate that is inevitably in store for most, if not all, of the Lanai native birds.[139]

The habits of this bird are very like those of the other species: short flights, finding its food in the bark of tree trunks and branches, greeting the stranger with its chipping call, but using it more sparingly than the Kauai or Molokai species...[140]

—George C. Munro, 1944

J. G. Keulemans del. et. lith Mintern Bros imp.

OREOMYZA MONTANA, (WILS.)

(1) Adult, (2) Juvenile

F. W. Frohawk del. et lith.

West, Newman imp.

1. LOXOPS COCCINEA
2. LOXOPS FLAMMEA

Kākāwahie (Molokaʻi ʻAlauahio)

Paroreomyza flammea

Description:

Adult male. Front and sides of the head pure scarlet; top of the head and back brownish scarlet, brightening into nearly pure scarlet on the rump; chin, throat, and lower surface generally pure scarlet, but paler in hue, brightening, however, on the flanks; remiges and rectrices blackish brown edged with brownish scarlet; wing lining pale scarlet. Bill and legs light pinkish brown.[141]

Adult female. Top of the head hair-brown, but each feather brownish scarlet at the base, and the shafts of those towards the back of the head grey; back hair-brown tinged with red; rump distinctly russet; upper tail-coverts brownish scarlet; remiges and rectrices blackish brown edged with brownish scarlet, as also are the upper wing-coverts. Beneath, dull white tinged with pale scarlet; sides of the body reddish brown, and wing-lining white tinged with scarlet.[142]

Measurements:

Total length 5 inches, wing from carpal joint 2.5, culmen .5, tarsus .75, tail 2.[143]

—Scott B. Wilson, 1899

Geographic range: Island of Molokaʻi.

Specimens in museums: 138.

Historical information: First specimen collected June 1880. Last observation May 1963.

Reports and observations:

It is peculiar, as far as I can judge, to the island of Molokai, and I only procured three specimens during my stay, all at Kalae; it may not, however, be safe to consider it rare, as my host easily obtained the native name for me, thus showing the bird to be known to the aborigines. I met with all three examples on the same day, killing a male and female at the same shot.[144]

While we were wandering about and searching for the trail, I heard a curious sound,—a continued chip, chip, chip, not unlike the sound of chopping wood when heard at a distance—which at first I did not think could belong to a bird; soon, however, I was undeceived, as a flash of brilliant orange colour passed us in the fog; when, on trying to follow it up, the continuous metallic note enabled me to get within range and I fired, bringing down two birds, which proved to be male and female. Soon afterwards I shot another of the bright-coloured males.[145]

—Scott B. Wilson, June 1890

J. G. Keulemans del. et. lith Mintern Bros imp.

OREOMYZA FLAMMA (WILSON)

(1) Male, (2) Female, (3) Male Juvenile, (4) Female Juvenile

My collector found this remarkably beautiful bird quite common in all the upper regions of Molokai, chiefly in the districts of Pukoo, Alawa, and Kalae. The birds were met with almost everywhere in the forests, from the deeper valleys up to highest peaks, and were not at all rare, but were never noticed among the Kukui trees at the bottom of the ravines. They are fairly active and seemed to have no preference for any particular kind of tree, but were perhaps more often seen on the Ohia trees.[146]

They were not at all shy, but, on the contrary, rather inquisitive, and could without difficulty be attracted when their call-note was properly imitated. Their food consists of insects.[147]

—Walter Rothschild, 1900

In habits the Kakawahia resembles the species of the genus to which it belongs, and from which it differs in color so widely. They prefer to feed over the trunks and branches of the trees. Here they secure the insects that make up almost the whole of their diet. However, they will be seen in the tops of the tallest trees, but apparently paying little or no attention to their flowers. In short, they are persistent and sturdy entomologists, always active and alert, but strange to say they, seldom if ever, take insects on the wing.[148]

When they have once settled on a home in the forest they at once set about to establish their sphere of influence, over which they rule, so as far as possible, to the excursion of every trespasser. On the approach of some intruder, as a man or dog, they will both set up a scolding "Chirk, chirk", that is no uncertain sound to one familiar with birds voices and ways.[149]

—William A. Bryan, 1907

Endemic to Molokai, and in danger of extinction, in the 1890's this beautiful bird was found in the upper boggy forest. I collected a series of specimens in 1907 when it was still quite common.[150]

There is little hope of saving this interesting and beautiful bird, unless it has a refuge, as I hope it has, with other Molokai native birds on the elevated plateau between Wailau and Pelekunu Valleys.[151]

—George C. Munro, 1944

"The last records of this species were by N. Pekelo, who describes two birds near Pepeopae Peak 21 May 1961, three at an unknown locality 8 Jun 1962, two heard and one seen at unknown localities 29 Apr 1963, and two pair between Waikolo and Pelekunu valleys in late Apr-early Jun (probably May) 1963 (Pekelo 1963, 1964). The combination of plumage and behavioral notes are sufficient to rule out other Moloka'i birds and we consider these as the last substantiated records of the species; a mention for a sighting as late as 1967 (E 37:107) appears to be in error."

"Searches in 1972 (Pratt 1973), 1975 (Scott et al. 1977), 1979-1980 (Scott et al. 1986), and 1988 and 1995 (P. E. Baker and Baker 2000, Reynolds and Snetsinger 2001) failed to turn up Kakawahie and we presume it is now extinct. Based on Poisson analyses of persistence probabilities using confirmed and unconfirmed records, Elphick et al. (2009) estimated that it went extinct in 1910-1969, with upper limits of 1920-1985; see also Scott et al. (2008). The Kakawahie was listed as endangered by the USFWS in 1970 and by the State of Hawaii in 1982 (USFWS 1984, 2006)."[152]

—R. L. Pyle and P. Pyle, 2017

F. W. Frohawk del. et lith. West, Newman imp.

HIMATIONE MACULATA

Male Juvenile

O'ahu 'Alauahio

Paroreomyza maculata

Description:

Adult male. Very similar to the adult male of *H. chloris*, but with the olive upper plumage darker, though tinged with yellow; forehead hardly brighter than the crown, but a distinct, though indefinitely marked, yellowish streak over the eye; lores brownish black; chin, cheeks, auriculars, and throat clear golden-yellow, which colour pervades the breast and belly, becoming very pale, almost white, on the abdomen; lower tail-coverts pale yellow. Wing-coverts with distinct whitish marks of considerable size.[153]

Adult female. Very unlike the male above described. Streak over the eye and under parts white, tinged with yellow; sides of the body and flank-region grayish upper parts olive-grey, showing whitish marks, which are much less distinct than in the male. Bill and feet rather dark brown. "Soles flesh colour and orange; iris dark brown (Rothschild)."

Measurements:

Total length about 5 inches, culmen about .6, wing 2.6-2.8, tail nearly 2, tarsus 8.[154]

—Scott B. Wilson, 1899

Geographic range: Island of O'ahu.

Specimens in museums: 120.

Historical information: First specimen collected 1836-1837. Last observation 1968.

Reports and observations:

This straight-billed bird, found only upon Oahu, was described by Prof. Cabanis, in the 'Museum Heineanum,' from a male and an immature female obtained in that island by Deppe when in company with Townsend in 1836-37.[155]

Himatione maculata is fairly common in the district of Halemanu, where there are still some remains of the former forest; and Palmer found it "not rare" in the upland region of Waialua at an altitude of 1500 feet and upwards, while Mr. Perkins obtained a considerable number of specimens at the same place and at Kawailoa in 1893, some of them at a rather lower elevation.[156]

—Scott B. Wilson, 1899

Only on April 10th (1899) the first specimen of Akikiki *(Oreomyza masculata)* (Cab.) was seen and shot, and which a fine series was brought together during the rest of the stay on Oahu.[157]

—Henry C. Palmer, 1899

The Oahuan *O. maculata* is less numerous than any of these, though a common enough species, and is found on both mountain ranges, but it seems to have entirely disappeared from the mountains in the immediate neighbourhood of Honolulu, where it formerly occurred.[158]

—R. C. L. Perkins, 1903

It was fairly common in the 1890's. I tramped many miles of newly made C.C.C. trails on Oahu in 1935 and did not see a single individual but other observers report having seen it since, commonly, thus promising a prospect of survival.[159]

—George C. Munro, 1944

"The last substantiated records of the O'ahu 'Alauahio were of specimens collected by Loye H. Miller at Leilehua Ranch in the Wai'anae Mts above Wahiawa in Apr-May 1903, by L. Richards near Pu'u Kamana (between Waiawa and Kipapa streams) at 425 m elevation 20 Jul 1950, and by G.E. Swedberg at 700 m elevation along the Poamoho Trail in the nc. Ko'olau range 19 Sep 1968 (not 9 Sep as reported E 38:136). The 1950 specimen (MVZ 122617) is a flat skin only (no skull) and the 1968 specimen (BPBM 155011) is preserved in fluid but identifications of both appear correct (PP examination). Flocks of 30-50 individuals were reported on the date the last specimen was secured but it seems likely that O'ahu 'Amakihis may have comprised most of this flock. Banko (1979, 1984) summarized early observations and localities of most specimens collected. It was listed as endangered by the USFWS in 1970 and by the State of Hawaii in 1982 (USFWS 2006). Since the last verified specimen was collected, in 1968, there have been more than 30 sight reports of O'ahu 'Alauahio, through 2002, from throughout the Ko'olau and Wai'anae ranges. This contrasts with the observations of Munro (1944, E 11:29), who hiked more than 200 miles throughout O'ahu in 1935 without observing a single individual, and W. Donagho (E 27:96-97), who failed to find any during his extensive forays between 1947 and 1966. Shallenberger and Pratt (1978) and Shallenberger and Vaughn (1978) vetted all records published in the 'Elepaio and elsewhere between 1940 and 1978, and considered only three records to be "virtually certain," the remainder ranging from "probable" to "highly unlikely" (see also Elphick et al. 2009). Some of these records include well-described birds (e.g., Shallenberger and Vaughn 1978; E 19:26, 35:31, 46:132-133); however, many published observations indicate uncertainty (cf. Shallenberger 1977) and several of these descriptions were based on birds "creeping" (which 'alauahio's are not known to do) and/or favor juvenile O'ahu 'Amakihis, which can be very similar in appearance. The vast majority of recent reports of O'ahu 'Alauahios are from Jul-Nov, when juvenile 'amakihis are prevalent. One individual identified as an O'ahu 'Alauahio was studied by numerous observers at a sap leak of a koa tree along Aiea Trail above Aiea in Dec 1976 and it was concluded after long evaluation to have been an O'ahu 'Amakihi. Because of the similarity between these two species, potential confusion with the Japanese Bush-Warbler, and the fact that field guides may have been misleading (cf. Shallenberger and Pratt 1978), we consider all sight reports since 1968 as unsubstantiated. Extensive searches during the 1990s (see P. Baker and Baker 2000, Pratt 2005) have failed to locate O'ahu 'Alauahio and, based on Poisson analyses of persistence probabilities using confirmed and unconfirmed records, Elphick et al. (2009) estimated that it went extinct (or will do so) in 1974-1981, with upper limits of 1989-2020; see also Roberts (2010). We consider the O'ahu 'Alauahio to be extinct."[160]

J. G. Keulemans del. et. lith — Mintern Bros imp.

OREOMYZA MACULATA, (CAB.)

(1) Male Adult, (2) Male Juvenile, (3) Female, (4) Pull.

—R. L. Pyle and P. Pyle, 2017

A Lāna'i Hookbill, *Dysmorodrepanis munroi,* preys on an extinct Lāna'i Tree Snail, *Partulina crassa*, a species wiped out by the introduced carnivorous Rosy Wolf Snail, *Euglandina rosea* (Spiraxidae), in the 1920s. The specialized jaws of the Lāna'i Hookbill may have been used to crush or to extract snails from their shells. Behind grows the critically endangered Lāna'i Lobelia, *Cyanea lobata baldwinii* (Campanulaceae), once thought extinct but now known from two plants discovered in 2006. (JULIAN P. HUME)

Lāna‘i Hookbill

Dysmorodrepanis munroi

Description:

General colour above grayish olivaceous, less grey on the middle of the back and on the rump; head above in front and an indefinite stripe above and extending behind the eyes much more yellow; underparts of neck and body entirely pale, appearing white, irregularly suffused with pale yellow. Wings with all the primaries except the outer one with very narrow but conspicuous yellowish margin outwardly, this colour becoming white (or hardly tinged with yellow) on the distal ends. The inner web is largely white in all these feathers basally, the white gradually extending on the inner primaries and in these reaching the tips of the feathers. The inner secondaries are broadly white-tipped and margined with white inwardly, the innermost ones with the whole inner web white. The breadth of the white tip decreases towards the primaries. The rectrices are dark, the outer margins narrowly olivaceous.[161]

Measurements:

Mr. Munro's notes give the following additions:—"Length six inches, sex not determined, the legs muscular with strong sinews, the jaw muscles more than usually developed, skull round almost like a marble, eyes large for the size of the bird, the iris dark brown, as also the upper mandible, the lower light brown, lighter beneath; legs light slate-colour, the soles of feet yellowish."[162]

—R. C. L. Perkins, 1919

Geographic range: Island of Lāna‘i.

Specimens in museums: 1.

Historical information: First specimen collected February 22, 1913. Last observation February 22, 1913.

Reports and observations:

I discovered this bird on February 22, 1913, in the Kaiholena Valley. It flew from a tree with a little chirp and alighted a short distance away, and thinking it was an ou I shot it. It measured 6 inches in length, larger than Pseudonestor of Maui and smaller than the ou.[163]

Scene from the native dry forest at Kaunolū, southwestern Lāna'i, in 1,000 AD. A trio of 'Ō'ū *Psittirostra psittacea* (center left) feed on the fruits of 'Ohe Makai *Polyscias sandwicensis*, soon to be joined by an 'I'iwi *Drepanis coccinea* (center right). A pair of Maui Nui 'Akialoa *Akialoa lanaiensis* (bottom left) forage on a tree trunk of Akoko *Euphorbia olowaluana*, along with a pair of Lāna'i Hookbill *Dysmorodrepanis munroi* (right) that sit beneath a fruiting branch. Depicted imaginary, the males of 'akialoa (bottom left) and hookbill (bottom center) probably acquired bright yellow underparts during the breeding season. 'Ohe makai and akoko trees formerly dominated the forests of Lāna'i. (JULIAN P. HUME)

Its striking peculiarity was its beak. Both mandibles were curved towards each other so that the tip of the lower is the only part that touched the upper leaving an opening between them like the Hawaii nukupuu, only more exaggerated.[164]

I saw a bird of this kind on two occasions afterwards, once on March 16, 1916 farther up the Kaiholena Valley, and again on August 12, 1918, at Waiakeakua at the southwest end of the forest. On both occasions I was sure that it was not one of the known birds of the Lanai forest.*[165]

It is my theory that this bird and the other rare Lanai bird, the Lanai akialoa *(Hemignathus obsurus lanaiensis)* both frequented the forest of akoko *(Euphorbia lorifolia)* that originally covered the plains now planted in pineapples. The akoko tree has fruit about the size of that of the opuhe *(Urera sandwichensis)* on which the bird was feeding when taken. As I understood from persons who remembered this forest it was a continuous area of thousands of acres of akoko trees.[166]

I tried for several years to obtain another specimen so as to allay any doubts that might arise that this specimen is a sport but failed to do so. From my experiences with this specimen and observations in the forest afterwards I am convinced that the species existed in the early years of this century.[167]

—George C. Munro, 1944

As so few specimens have been seen by so skilled a collector, the bird must be a great rarity, but its discoverer hopes that it may increase in numbers, as the forest is now rigidly protected and rapidly recovering. When I collected on Lanai in 1893 and subsequently the forest was in deplorable condition, being rapidly destroyed by countless wild goats, and it was also full of wild pigs and cats that had run wild. The latter, as I have elsewhere recorded, were destroying native birds wholesale. Only on the sheer sides of the mountain and on a very small part of the narrow backbone was the forest in a natural condition. Neither Lord Rothschild's collectors nor myself ever found a specimen of this bird.[168]

—R. C. L. Perkins, 1919

**(Please note these were not confirmed to species.)*

F. W. Frohawk del. ct lith. West, Newman imp.

VIRIDONIA SAGITTIROSTIS

Male and female

Greater ʻAmakihi

Viridonia sagittirostris

Description:

Adult male. Upper parts bright olive-green, rather brighter and more yellowish on the forehead, sides of the head, and upper tail-coverts. Underparts more yellowish olive-green; wings blackish brown, the primaries narrowly and the secondaries more broadly margined with yellowish green; tail blackish brown, with yellowish-green margins. Under surface of the wings dark ashy, the quills margined with dull white, tinged with greenish yellow. Iris hazel, with reddish tinge; upper mandible black, under mandible black with a brown tinge, both bluish slaty at base; tarsi bright blackish grey; soles flesh-colour with a yellow tinge.[169]

Adult female. The first female sent by Palmer, the type of the original description, is rather duller in tinge of colour both on the upper and under parts, and appears to be a younger specimen, for another female is entirely similar in colour to the male. The iris of the former was "dark hazel," much darker than in the adult male.[170]

Measurements:

ad.(type). Wing 3.25 inches, tail 2.1, tarsus 0.86, culmen 0.89 Total length, measured by Palmer in the flesh, 6¼ to 6½ inches.[171]

—Walter Rothschild, 1900

Geographic range: Island of Hawaiʻi.

Specimens in museums: 32.

Historical information: First specimen collected 1892. Last observation 1902.

Reports and observations:

Palmer describes the discovery of this rare and singular bird as follows:-April 23rd, 1892. When marching along the slopes of Mauna Kea, above Hilo, and not far from Wailuku, through old forest of Ohia, here and there intermixed with the Koa, and much enjoying the beautiful contrast of thousands of bright crimson-coloured flowers of the lofty Ohias, which were frequented by numerous birds of the common Hawaiian species, I heard a strange note coming from an Ohia tree that I had just passed. At the first moment I thought it might be an Akialoa's note, but on hearing it again was convinced I had never heard it before. I soon saw the bird among the leaves, but before I could fire it flew into the tree just above my head, where it began to call again, as if calling its mate, until I shot it. On picking it up I saw it was quite new to me, and I hope it may be new to science. Its call-note might be described as a high clear 'chirrup' uttered three or four times in quick succession at short intervals. After waiting a few minutes its mate came up, but was shy and escaped me. I waited on the spot for about an hour, but none came back; on returning to my camp, however, I was fortunate enough to shoot another specimen, to which I was attracted by its cry, and I think I heard two or three more which I could not get; but so many birds of various kinds were singing among the flowering trees that it was very difficult to distinguish their various notes from a distance.[172]

Palmer obtained altogether four specimens only of this apparently rare bird. They are obviously forest-birds, but not inhabitants of high elevations, for they were only observed at heights of about 500-1500 feet in Hilo district and all within a radius of a few miles. They were mostly seen in the higher trees, but one was once observed on a banana-leaf not ten feet from the ground.[173]

—Walter Rothschild, 1900

J. G. Keulemans del. et. lith Mintern Bros imp.
VIRIDONIA SAGITTIROSTRIS, ROTHSCH.
(1) Female, (2) Male

A few birds appear to have been so rare and so local that they were quite unknown to the natives; at all events they seem not to have been named by them. The Viridonia, which is confined to a very small area of dense forest on either side of the Wailuku river, Hawai'i, is a conspicuous example of this kind. The writer has shown skins of this bird to a number of natives reputed to be well aquainted with Hawaiian birds- one of whom was born within Viridonia's territory- but none of them had ever seen it alive or could give it a name. One of the old men, indeed, maintained its identity with the amakihi, although it is nearly twice the size of that bird and by no means of the same color.[174]

This beautiful bird was one of the most interesting discoveries made by Palmer, Mr. Rothchild's collector, in the island of Hawaii where alone it is found. It is one of the rarest prizes among existing Hawaiian birds…[175]

The long straight bill is oriole-like in character and, like the orioles, the Viridonia seems to live chiefly upon insects which it gleans from the foliage of the ohias, to which tree it seems to confine its attentions chiefly.[176]

—H. W. Henshaw, 1902

F. W. Frohawk del. et lith. West, Newman imp.

HEMIGNATHUS PROCERUS

Kaua'i 'Akialoa

Akialoa procerus

Description:

Adult male. Front and top of the head dark ashy olive, shading into olive on the back, which becomes brighter on the rump; sides of the face olive, with an indistinct olive line over the eye; body beneath sulphur-yellow, light primrose on abdomen, the flanks washed with olive; under tail-coverts olive-yellow; wing- and tail-quills ashy brown, edged outwardly with olive; secondaries and wing-coverts ashy brown, very broadly edged with olive; irises dark hazel; bill black; feet bluish black.[177]

Adult female. Above dingy yellowish buff, the feathers on the head being yellowish olive with black centres; the yellow mark over the eye more distinct, owing to the blacker lores; beneath uniform olive-buff. In both male and female the edge of wing and wing-lining is white, slightly tinged with primrose-yellow.[178]

Measurements:

Adult male. Total length 7.50 inches, wing from carpal joint 3.75, tarsus 1, tail 2, hind toe with claw .75, middle toe with claw .95, maxilla following the curve 2.80, chord subtending the curve 1.75, difference between maxilla and mandible .30.[179]

Adult female. Maxilla following the curve 2.55 inches, difference between maxilla and mandible .25.[180]

—Scott B. Wilson, 1899

Geographic range: Island of Kaua'i.

Specimens in museums: 102.

Historical information: First specimen collected ca. 1887. Last observation 1969.

Reports and observations:

This bird is much more common and enjoys a wider range than the Nukupuu, which bird it much resembles in habits. It seems to inhabit the whole forest-region of Kauai; its food consists of insects, their eggs and larvae, and we have also seen them sucking honey from the Lehua flowers. Above Makaweli in January and February we found it less common than at other places we visited: there they were mostly on the koa trees (being the most suitable hunting-ground for them in this locality). Usually there was a pair in the same vicinity, but not keeping very close together, so that when one was shot we would usually get another.[181]

At Kaholuamano in the latter end of February and beginning of March they were more common, generally, in company with the Akikiki, feeding on the Lehua trees, the pairs keeping more together. In one instance I shot a female, and the male stopped in the top of the tree calling desperately. I fired at him without effect, and so intent was he in looking for his mate that he immediately returned and was brought down by another shot.[182]

Their chirp seemed different here; Mr. Palmer likens both the chirp and song to that of the canary. We watched a pair singing together one day; the smaller and duller bird (probably the female) seemed to have fewer notes than the other.[183]

I have seen these birds from the branches in the tops to the roots of the trees, probing into holes and under the bark, where they find a harvest of cockroaches' eggs, beetles, and grubs; on one occasion I saw one alight on the ground and insert its bill amongst mats of dead leaves and bits of wood; have also seen them collecting insects from the bases of the leaves of the halapipi tree; have not often seen them feeding on honey. In feeding they do not seem to depend much on sight; have never noticed them to look into a crevice, as the A-A, before inserting their bill. I saw one send its bill at full length into a hole in a tree; have seen them work about one spot for some minutes, but have not noticed them break off any portions of bark or wood. Like the Nukupuu, it is an active bird but can be easily approached within gunshot with ordinary caution. Have also a strong smell when killed; and some, shot at Makaweli, had sores on their feet like the other birds in that locality at that time.[184]

—Scott B. Wilson, 1899

J. G. Keulemans del. et. lith Mintern Bros imp.

HEMIGNATHUS PROCERUS, CAB.

(1) Male adult, (2) Juvenile, (3) Female

The long-billed *Hemignathus* was seen by Henry C. Palmer on the south and south-west sides of Kauai at altitudes from 900 to probably between 3,000 and 4,000 feet above the sea. On the west side it was common at 3,000 feet, and only a few were seen higher. It was also met with on the north and north-east sides at elevations of about 600 to 2000 and 3000 feet, though not quite so numerous. Palmer believes it inhabits all the upland forests of the island.[185]

—Walter Rothschild, 1900

In all my visits to the Kauai forests since 1920 I have not seen or heard anything of this bird. Donagho reported seeing it on the upper plateau in October 1941. It is hoped his observation is correct and that the bird will survive.[186]

—George C. Munro, 1944

Only two individuals of this unique species were seen, both in high (4,000 feet), forested ridge country within a mile southeast of the upper Koaie River Cabin. …The specimen collected July 20 was a female apparently in post breeding condition. Among its stomach contents were: parts of four or more large beetles, probably carabids; parts of three or more large beetle larvae, probably a wood-boring species; two large black, hooked mandibles of beetle or spider; and one sprouting seed.[187]

—Frank Richardson and John Bowles, 1960

…last apparently valid sighting was in 1969 by Phillip L. Bruner.[188]

—R. L. Pyle and P. Pyle, 2017

Only two museum skins of the O'ahu 'Akialoa *Akialoa ellisiana* survive, both in a sombre, dull-green non-breeding plumage. As exhibited in the males of the closely related *Hemignathus*, this reconstruction shows the male in a probable brighter breeding plumage with deep yellow eye-stripe and underparts. (JULIAN P. HUME)

O‘ahu ‘Akialoa

Akialoa ellisiana

Description:

Description condensed from Lichtenstein (ut supra, p. 450).— Uniform olive-green above and on the wing-feathers, though the inner and concealed parts of the latter are dull brown. Beneath paler though as little brilliant. The chin, middle of the belly, and the lower tail-coverts pale cream-colour. A pale yellow stripe over the eye is enlivened by a dark brown streak running immediately beneath it from there to the bill.[189]

Measurements

The lower mandible is 3 lines shorter than the upper. The whole length of the bird is 7 inches, of which the bill and tail measure each 1¾; the tarsus 11 lines; the middle toe with its claw 9 lines; the outer toe one half and the inner a whole line shorter than the middle toe.[190]

—Scott B. Wilson, 1899

Geographic range: Island of O‘ahu.

Specimens in museums: 3.

Historical information: First specimen collected 1837.
Last observed (specimen shot but never found) 1892.

Reports and observations:

Palmer did not find this species on Oahu, where it was discovered by Deppe; and having no specimen of it before me at present, I insert here a translation of Lichtenstein's original description. It runs as follows:—

Latham describes in the genus *Certhia* a species from the Sandwich Islands, which struck him particularly in so far that the lower mandible was a quarter of an inch shorter than the upper. This species was included in the system under the name of *Certhia obscura;* but it is so distinct from all the different subgenera into which this group has rightly been divided, that one is forced to make a new genus for it, for which I propose the above name (*Hemignathus*).[191]

J. G. Keulemans del. et. lith Mintern Bros imp.
HEMIGNATHUS ELLISIANUS, (GRAY.)

The most characteristic feature of this genus is certainly the unusual length and bow shaped curve of the bill and the unequal length of the mandible. But, moreover, we find in the great development of the hind limbs (the thickness of the tarsi, and the length of the toes and claws) differences from the allied groups which cannot be overlooked. The nearest ally of this new genus is *Drepanis* (Melithreptus) vestiaria. But the beak (irrespective of the length) is much less thick and robust at the base; the nostrils also are much narrower and hardly feathered at the base, while in *Drepanis* (Vestiaria) they are feathered for their whole length.[192]

As neither Palmer, Perkins, nor Wilson found this bird, there is every reason to believe that it is extinct. On the other hand, collecting in the dense forest and rugged mountains of Oahu is so difficult that—as Palmer himself says—one might be months or even years busy in collecting and yet fail to find a very rare bird.[193]

—Walter Rothschild, 1900

F. W. Frohawk del. et lith. West, Newman imp.
HEMIGNATHUS LICHTENSTEINI

Maui Nui ʻAkialoa

Akialoa lanaiensis

Description:

Male ad. (sexed by H. C. Palmer). Above yellowish olive-green, somewhat mixed with grayish brown on the head where the brown bases show through. Quills dark brown, edged with the colour of the back; rectrices same. Below yellowish olive-green, much less bright than on the back and with little yellow in it, shading into olive-buff on the vent, and with a brown pale shade on the chin and throat; under wing coverts pale buff, with an olive-yellow tint on the edges of the feathers; feathers on the tibia buffy white.[194]

Female ad. (sexed by H. C. P.; ovaries enlarged). Above more greenish olive and less bright; the superciliary stripe faint and greenish; chin, throat, and middle of abdomen buffish yellow; yellowish olive-green on the sides of the body and breast. In other respects similar to the male in colour.[195]

Measurements:

Male. Palmer gives the total length as about 7½ inches, but it is only 6 inches in the skin; wing 3.3, tail 2.1, tarsus 1.93, middle toe without claw 0.7, culmen along the ridge 2.1 (given as 3.1 but a slip of the pen in the original description, l.c.), bill from base to tip in a straight line 1.9 inch.[196]

Female. Wing 3.05 inches, tail 1.9, culmen 1.9, bill from base to tip in a straight line 1.7.[197]

—Walter Rothschild, 1900

Geographic range: Islands of Lānaʻi, Molokaʻi and Maui.

Specimens in museums: 3.

Historical information: First specimen collected November 15, 1892. Last observation January 9, 1894.

Reports and observations:

Palmer found this rare bird in the mountains of Lanai, where it inhabits the upper forest region. It seems to live on insects which it extracts from the partially decayed bark of the various trees. The stomachs of the three specimens which were procured were full of beetles and other insects. They were also seen sucking on the flowers of the "Haha." Palmer found them shy and very rare. They were very quick in their movements from flower to flower. Their flight is quick and straight; no jerky movements were noticed.[198]

Palmer heard the birds utter "a whistle, clear, but yet with something hollow in it, very difficult to express in words." Their note and song resemble those of *Hemignathus obscure*s from Hawaii, but are much clearer.[199]

The greater portion of Lanai is barren and the island in general suffers from the long droughts which frequently prevail; the ravines and glens, notwithstanding, are filled with thickets of trees. According to Palmer the trees are nowhere very lofty, and the forest seems to be dying out. The cause is attributed to the great number of goats and cattle that eat up the undergrowth.[200]

Bird-life is very scarce, according to Palmer, and he thinks the native birds are dying out fast, partly on account of the supposed disappearance of the forest, partly from the great number of introduced cats. Unfortunately, some introduced birds, as the Carpodacus and the Rice-birds, were found in numbers, certainly not to the advantage of the native birds.[201]

—Walter Rothschild, 1900

J. G. Keulemans del. et. lith Mintern Bros imp.

HEMIGNATHUS LANAIENSIS, ROTHSCH.

(1) Female, (2) Male, (3) Male juvenile

F. W. Frohawk del. et lith. West, Newman imp.

HEMIGNATHUS OBSCURUS

Lesser ‘Akialoa

Akialoa obscurus

Description:

Adult male. Head and upper parts generally uniform dull greenish olive, rather brighter on the rump; lores dusky, with a yellow mark over the eye; throat, sides of face, and breast dull olive-green, lighter on abdomen and under tail coverts; wing- and tail-quills ashy brown, edged with dull olive; irides dark hazel; bill and feet dark brown.[202]

Adult female. Nearly similar to the male, but perhaps rather duller in plumage.

Measurements:

Adult male. Total length 5.50 inches, wing from carpal joint 3.5, maxilla following the curve 1.85, chord subtending the curve 1.50, difference between maxilla and mandible .15, tarsus .85, middle toe with claw .80, hind toe with claw .75, tail 1.60.[203]

Adult female. Total length 5.15 inches, wing from carpal joint 3, maxilla following the curve 1.45, chord subtending the curve 1.30.[204]

—Scott B. Wilson, 1899

Geographic range: Island of Hawai‘i.

Specimens in museums: 106.

Historical information: First specimen collected 1779.
Last observation 1903.

Reports and observations:

… entirely green with a tinge of yellow [and is called akaiearooa].[205]

—Captain Charles Clerke, 1779

This species—peculiar, so far as my observations go, to the Island of Hawaii occupies the lower forest-zone from about 1100 to 2500 feet, and is most plentiful among the tall ohia trees. Like its larger relative *H. stejnegeri* on Kauai,—it prefers decayed timber in which to search for its food, and invariably chooses a rotten or half dead tree for its hunting-ground, no doubt on account of its slender bill, which requires soft material to work upon. It is also very partial to the great tree-ferns which in the forests of Hawaii reach a height of more than 30 feet, and, as the sombre colour of its plumage is very nearly that of their foliage, it is most difficult to observe, and is at the same time more quiet and unobtrusive in its habits than any other member of the genus; in fact, had it not been for its clear and characteristic call-note, I doubt whether I should have noticed it at all. It must—at least in the several localities I visited and at the time of year I saw them—be considered a scarce bird: and whilst I was at Olaa in the district of Puna—a place renowned in ancient times for its bird-catchers an old native, Hawelu, an excellent observer and well skilled in the almost forgotten art, told me that it was extremely rare. During a long stay in the higher forest region in Kona, I did not notice it, and believe, as I remarked above, that it is confined to the lower forest-zone.[206]

—Scott B. Wilson, 1899

Palmer believes that these birds inhabit pretty well all the upland forests on Hawaii. He found them common in the district of Kona, on the south-west side of the island, at altitudes of about 2000 to 6000 feet. He saw none in Kohala district (north-side) or Kohala mountain. They were comparatively numerous in Hamakua district, north-east, between 2000 and 5000 feet. Numbers were seen in Hilo district above Hilo, at heights of 1400 to 3000 feet, and a few in the upper parts of Puna district on the east side.[207]

They inhabit dense forest, where they generally were not at all rare, although other birds were more numerous. Their principal food consists of insects, which they capture in the bark and decaying stems and branches. In collecting their food they were seen to put their bills into the crevices of the bark up to the base.[208]

—Walter Rothschild, 1900

J. G. Keulemans del. et. lith Mintern Bros imp.
HETERORHYNCHUS OBSCURUS, (GM.)
Male and female

Hemignathus obscurus has been found in each district of the large island and frequents both the lower and upper parts of the continuous forest. It was common in Kona in 1892 and 1894, not only in the denser forest but also in the open parts at all elevations, abundant in Kau district in the Koa woods three or four miles from Kilauea, numerous in the heart of the virgin forest of Mauna Kea, north of the Wailuku river, as also in the Kohala mountains. In Olaa it was less common, at least at 1500 to 2000 ft., where I chiefly observed it; and here the magnificent virgin forest has now been cut down for the purpose of planting coffee and sugar-cane.[209]

In the heart of the virgin forest on Mauna Kea, during three weeks of continuous rain in the winter months I daily observed the Akialoa visiting these flowers; as also on the rough lava-flows in the Kona district on the other side of the island. When visiting the Haha, the bird is often exceptionally tame, and with a little care all its movements can be watched from a distance of only a few paces, and when it is only a few feet from the ground.[210]

In Kona *H. obscurus* was common in the very tallest flowering trees, frequently two to eight in a single tree, but requiring to be looked for amongst the extraordinary numbers of commoner *Drepanids*.[211]

—R. C. L. Perkins, 1903

F. W. Frohawk del. et lith. West, Newman imp.

HEMIGNATHUS HANAPEPE
Female and male

Kaua'i Nukupu'u

Hemignathus hanapepe

Description:

Adult male. Front and top of the head dull gamboge-yellow, passing into bright olive, which extends over all the upper surface of the back, wing-coverts, and tail; lores black, joined by a narrow black line just above the bill; throat, cheeks, and breast deep gamboge-yellow, passing into dusky white on abdomen and tail-coverts; wing- and tail-quills greyish brown, edged outwardly with olive; irides dark hazel; bill and feet slaty black.

Adult female. Tinged with olive on the forehead, an indistinct line passing over the eye; wing- and tail-quills margined with olive; breast primrose-yellow, changing into dull white on the abdomen; lower tail-coverts tinged with buff.

Measurements:

Adult male. Total length 5.60 inches, wing from carpal joint 3.25, maxilla following the curve 1.20, chord subtending the curve .80, difference between maxilla and mandible .50, tarsus .90, middle toe with claw .70, hind toe with claw .75. tail 1.85.[212]

—Scott B. Wilson, 1899

Geographic range: Island of Kaua'i.

Specimens in museums: 34.

Historical information: First specimen collected 1888. Last observed 1899.

Reports and observations:

I discovered the Nukupuu whilst staying at a little mountain-cottage belonging to the Sinclair family in the higher forest-region of Kauai, at an altitude of some 3000 feet, to which excellent collecting-ground Mr. Aubrey Robinson most kindly accompanied me.

The first I shot, a fine male which was in a lofty ohia tree, I took to be *Himatione parva*, the brilliant yellow of the breast in both species being very noticeable, and I was therefore greatly delighted to find, on picking it up, that I had secured a form quite new to me.

J. G. Keulemans del. et. lith Mintern Bros imp.

HETERORHYNCHUS HANAPEPE (WILS.)

(1) Male adult, (2) Female juvenile

I agree with Mr. Palmer in believing that the bird merely inserts its long hooked beak into crevices and holes in decayed wood, extracting by that means the grubs and insects which abound under the bark; its habit of keeping along the upper surface of a branch and examining the sides within its reach we both noted. The food doubtless consists for the most part of insects, larvae, small beetles, &c, but I am assured by my native—Keawe—that the Nukupuu also feeds on bananas and oranges, and I have every confidence in the assertion[213]

—Scott B. Wilson, 1899

Palmer found this species very scarce. All he procured were collected on the southern side of Kauai—two above Makaweli at a height of about 1000 feet, the rest at elevations of 3000 feet or so, in the district of Waimea, all within a range of fifteen miles. None were seen elsewhere. They kept singly or in pairs, and were almost invariably seen in company with *Oreomyza bairdi* (Akikiki).

These birds were never observed among the foliage, but were always busy on the bark of the trees, in the crevices of which they obtained their food. They were pretty lively in their movements, but not at all shy, in fact less so than many other birds. Their food consists of beetles and other insects, caterpillars, grubs, etc., which are usually hooked out of crevices and holes in the bark and decaying trunks and branches.[214]

—Walter Rothschild, 1900

I took a specimen in December 1898, and two more on May 6, 1899, all males. I doubt if it has been seen since. It was tame and unafraid, not seeming to notice our approach.[215]

—George C. Munro, 1944

"Bryan and Seale (1901), Munro (1944), and others (e.g., E 27:45, 47) failed to locate the Kaua'i Nukupu'u despite extensive searches. Beginning with a "fleeting glimpse" of two birds on 16 Aug 1960 (Richardson and Bowles 1964), there have been at least 26 sight records of Kaua'i Nukupu'u in the upland forests of Kaua'i through Jan 1996 (e.g., E 22:3, 26:29, 29:20, 37:107; Conant et al. 1998), although Sincock et al. (USFWS 1983c) did not record any confirmed observations in over 500 days of extensive fieldwork in 1968-1975. We follow Pratt and Pyle (2000) in considering these unsubstantiated, and we presume it went extinct in the early 20th century. Confusion with the similar 'amakihis, which can sometimes have abnormal bills, seems to be the primary source of misidentifications of nukupu'us on both Kaua'i and Maui (Pratt and Pyle 2000; also, cf. Olson and James 1994 regarding the interpretations of Cassin). Based on Poisson analyses of persistence probabilities using confirmed and unconfirmed records, Elphick et al. (2009) estimated that the Kaua'i and/or Maui nukupu'us went extinct in 1901, with an upper limit of 1906; see also Scott et al. (2008) and Roberts et al. (2010). Although proclaimed "extinct" by the USFWS in 1966 (Banko 1968), the Kaua'i Nukupu'u was subsequently listed as endangered by the USFWS in 1967 and by the State of Hawaii in 1982, and it continues to be so listed in the extremely remote chance that it is extant (USFWS 1983c, 1984, 2006)."[216]

—R. L. Pyle and P. Pyle, 2017

F. W. Frohawk del. et lith.

West, Newman imp.

HEMIGNATHUS LUCIDUS

O‘ahu Nukupu‘u

Hemignathus lucidus

Description:

Adult male (in the Paris Museum). Above olive-green, darker and more olive on the back, lighter and more green on the head, wing, and tail coverts. Lores and a line behind the eye brownish black; across the forehead a narrow and above the eyes a conspicuous orange-yellow superciliary stripe. Quills deep brown, outer webs edged with greenish yellow. Chin, throat, and upper breast bright orange-yellow; abdomen yellow and fading into pale greenish grey on the vent and under tail- coverts.[217]

Adult female or *immature male* (in the Senckenbergische Museum at Frankfort). Above dull brownish olive, tinged with greenish on the top of the head, rump, and upper tail-coverts, and on the edges of the quills and tail-feathers. Lores dusky; a somewhat ill-defined but distinct superciliary stripe; sides of head and throat yellowish. Rest of underparts light grayish brown, brownish olive along the sides of the body and flanks.[218]

Measurements:

The Frankfort specimen measures:-Total length 5½ inches, culmen 1.1, length of under mandible 0.7, wing 2.95, tarsus 0.76, tail 2.9.[219]

—Walter Rothschild, 1900

Geographic range: Island of O‘ahu.

Specimens in museums: 9.

Historical information: First specimen collected 1837. Last observed 1841.

Reports and observations:

Our knowledge of this species, as of *H. lichtensteini,* is due to the Prussian collector Deppe, who sent specimens of it which he obtained in Oahu to the Museum of Berlin, where they were described and figured by its Director, the celebrated Lichtenstein, as above stated. While Deppe was in the Sandwich Islands he was joined by the still better-known American naturalist Townsend, who (with Nuttall) accompanied Captain Wyeth's expedition across the Rocky Mountains to the Columbia River, and thence proceeded to Honolulu. In his work Townsend states that he and Deppe, in January 1837, hired a house in the Nuano [Nuuanu] Valley, five miles from that town, with the object of collecting birds and plants, and we may well suppose that this species (and *H. Lichtensteini* also) was found by them in that district. On Townsend's return he sent several specimens of birds collected by him to Audubon, then in this country, and among them two of the present species, which were acquired by the late Sir William Jardine, at the sale of whose collection, in 1886, they were bought for the Museum of the University of Cambridge.[220]

—Scott B. Wilson, 1899

J. G. Keulemans del. et. lith Mintern Bros imp.

HETERORHYNCHUS LUCIDUS (LICHT.)

(1) Male adult, (2) Male juvenile, (3) Female

Palmer was told that it used to be not uncommon just above Honolulu but no recent collector has been able to find this bird. It is evidently extinct and one of the rarest birds in collections.[221]

—Walter Rothschild, 1900

Of the four known species the Oahuan *H. lucidus* is almost certainly extinct, no individual having been obtained for many years. From such information as I have been able to gather from native or other sources it occurred in some numbers at least until about 1860, and frequented what must now be considered as the lower belt of continuous forest, its range in fact being apparently that of the Koa tree. Over the greater part of Oahu the Koa fails long before the main ridge of the mountains is reached, and an examination of the lower edge of the existing forest shows that a large, and probably much the larger part of the Koa belt has now been destroyed. If the bird was really restricted to this lower forest I should consider its total extinction as almost certain. The specimens obtained by Deppe and Townsend in 1837 were from Nuuanu valley, at no great elevation above the sea and were said to frequent the flowers of the banana. No doubt at that time the valley was densely forested for the greater part of its length, since Bloxam records that in 1825 he visited its head by a narrow path through a dense and shady forest. Lying as it does immediately behind the city of Honolulu this fine valley is now practically denuded of its forest and largely filled with unsightly guava scrub; and the cliffs on either side, which to Bloxam appeared bare in comparison with the valley itself, now alone retain their native vegetation.[222]

—R. C. L. Perkins, 1903

F. W. Frohawk del. et lith. | West, Newman imp.

HEMIGNATHUS AFFINIS

Maui Nukupu'u

Hemignathus affinis

Description:

Adult male. Top of the head bright gamboges-yellow, almost golden yellow, passing into yellowish green on the nape and hind neck, which colour is not gradually, but somewhat abruptly, terminated by that of the back, which is grayish olive-green without yellow, becoming brighter and greener on the upper tail coverts. Quills and rectrices deep brown, bordered with bright olive-green. Lores black, connected by a black band across the forehead, which is distinctly broader than in *H. Hanapepe*, Wils. A very tiny black chin spot. Throat bright yellow, gradually passing into a sulphur-yellow on the abdomen and pale yellow on the vent and under tail-coverts. Sides of the body tinged with greenish olive. Iris dark brown. Bill blackish brown, slate-colour at base. Tarsi and feet dark grey with a yellowish tinge.[223]

Adult female. Quite different from the corresponding sex of *H. Hanapepe*, Wils. Above olive-green with grayish shade, brighter and more green on the upper tail-coverts; on the forehead from the base of the upper mandible runs a bright yellow line to just above the eyes. Lores dusky. Sides of the head yellowish. Chin and throat bright yellow: middle of breast and abdomen pale yellow, much like "Naple's yellow"; sides of breast and flanks washed with olive ,under tail coverts slightly so. Under wing-coverts yellowish white.[224]

Measurements:

Male-Wing 3 inches, Tail 2, Tarsus 0.75, Culmen 1.19, upper mandible from the base to the tip 0.66[225]

—Walter Rothschild, 1900

Geographic range: Island of Maui.

Historical information: First specimen collected 1892. Last observed 1902.

Specimens in museums: 36.

J. G. Keulemans del. et. lith Mintern Bros imp.

HETERORHYNCHUS AFFINIS, (ROTHSCH.)

(1) Male, (2) Female

Reports and observations:

This species was discovered by my collector in 1892. It inhabits the forest-region of the little-known and unexplored island of Maui. Although not at all plentiful it was less rare than *H. Hanapepe* on Kauai, to which it is very similar in habits. It is not shy, but somewhat scarce. The stomach contained small beetles, skins of large grubs, and other insects. Sometimes this species gives vent to a prolonged song, almost similar to that of the introduced American Carpodacus, though considerably lower in tone. Its call-note is like that of *H. wilsoni* from Hawaii.[226]

As regards its mode of feeding, Palmer says:-"It uses its lower mandible to hammer away at the bark and then hooks the insects out with its long upper mandible. The hammering sound can be heard at a short distance. It also pulls insects out of the moss with the upper mandible and then, to kill them, shakes them like a dog does a rat."[227]

—Walter Rothschild, 1900

"Munro (in Gregory 1929) and Richards and Baldwin (1953) failed to find Maui Nukupu'us despite careful searches. Pratt and Pyle (2000) noted an additional 27 sight records between 1967 and 1996 (e.g., Banko 1968, Conant and Stemmermann 1980, Scott et al. 1986, Reynolds and Snetsinger 2001; E 37:107, 41:59, 41:77, 44:6, 50:68; HFW 9[3]:3) which they considered unsubstantiated. Confusion with the similar 'amakihis, which can sometimes have abnormal bills, seems to be the primary source of misidentifications of nukupu'us (Pratt and Pyle 2000; also, cf. Olson and James 1994 regarding the interpretations of Cassin). Based on Poisson analyses of persistence probabilities using confirmed and unconfirmed records, Elphick et al. (2009) estimated that the Kaua'i and/or Maui nukupu'us went extinct in 1901, with an upper limit of 1906; see also Scott et al. (2008) and Roberts et al. (2010). Although proclaimed "extinct" by the USFWS in 1966 (Banko 1968), the Maui Nukupu'u was subsequently listed as endangered by the USFWS in 1967 and by the State of Hawaii in 1982, and it continues to be so listed (USFWS 1983c, 1984d, 2006). We consider the Maui Nukupu'u extinct. There are no substantiated records of nukupu'us for the island of Hawai'i, although a much larger congener is known from the fossil record (James and Olson 1991, 2003; Olson and James 1994b; Burney et al. 2001; Pratt et al. 2001). A specimen collected during the U.S. Exploring Expedition in 1840-1841, along with Peale's (1848) account of "Hemignathus lucidus" on Hawai'i, caused Olson and James (1994) to speculate on the species' historical presence there, but the specimen turned out to be of the O'ahu subspecies (R. Fleischer in Pratt et al.2001, James and Olson 2003). van Riper (1982a, 1982b) noted a bird he thought to be a nukupu'u in the Kohala Mts 29 Jun 1971 (originally considered an Akiapola'au; E 34:1), but as with other 20th-century sightings (T. K. Pratt and Pyle 2000), we consider this record unsubstantiated, perhaps a Hawai'i 'Amakihi with a deformed bill."[228]

—R. L. Pyle and P. Pyle, 2017

O‘ahu ‘Ākepa

Loxops wolstenholmei

Description:

Adult male (shot on Oahu by Wolstenholme; type of *L. wolsenholmei*) Above dark rufous orange with a brownish wash, becoming lighter and brighter on the lower rump and upper tail-coverts. Wings blackish brown, edged with greenish orange on the outer webs and with pale ashy on the inner webs. Tail-feathers paler than the wings, and edged with bright orange on the outer webs. Underparts reddish orange, changing into bright salmon-orange on the lower abdomen and under tail-coverts. Under wing-coverts whitish grey, slightly washed with rusty brown. Iris dark hazel; maxilla whitish blue with dark brown tip, mandible the same; feet and toes black, soles light flesh-colour.[229]

Measurements:

Palmer measured the type in the flesh as 4⅜ inches; In skin it measures nearly 4 inches. Wing 2.2. tail 1.7, tarsus 0.74, culmen 0.45.[230]

—Walter Rothschild, 1900

Geographic range: Island of O‘ahu.

Historical information: First specimen collected 1825. Last observed 1900 or 1901.

Specimens in museums: 20.

Reports and observations:

Besides this perfectly adult male, in my museum at Tring, I am aware of specimens of adult males in the following collections:

Three in the British Museum, namely Bloxam's types and one from the Gould Collection, probably the one figured in the 'Voyage of the Blonde.'

One in the Berlin Museum, collected by F. Deppe on Oahu.

One or more in Cassel, collected by Deppe on Oahu.

One in Count Berlepsch's collection, also from Deppe, collected on Oahu.

One in Vienna, collected also by Deppe on Oahu.[231]

F. W. Frohawk del. et lith. West, Newman imp.

LOXOPS RUFA

It inhabits Oahu and is a very distinct form. It was, no doubt, more numerous in former days. At present it must be decidedly rare, as only one male was obtained in the course of many months. It was shot on April 24th (1893) in the mountains of the Wailua district, Oahu, by Palmer's assistant Wolstenholme. Mr. Perkins was present when he shot it, and neither he nor anyone else has succeeded in observing another example.[232]

—Walter L. Rothschild, 1900

F. W. Frohawk del. et lith. West, Newman imp.

LOXOPS AUREA

Maui 'Ākepa

Loxops ochraceus

Description:

Adult male. Dark orange ochraceous above and below, brighter and more yellowish towards the rump and upper tail-coverts above and towards the vent and under tail coverts below. Wing-quills blackish dusky, outer webs bordered with orange-yellow, inner webs with pale cinereous. Rectrices deep dusky brown, outer webs edged with orange-yellow. Under wing-coverts yellowish. Iris dark hazel; beak slate-blue, tip dark brown. Tarsi and toes greyish brown. Size not perceptibly different from that of *L. coccinea*.[233]

Other males shot at the same time of the year are wax-yellow or gall-stone-yellow. I am inclined to think that they are not so old as the orange-ochraceous ones, but Mr. Perkins, although admitting "occasional intermediate forms," is of opinion that the male is dimorphic in colour.[234]

Adult female. Above dark green, lighter and more yellowish green on the rump and upper tail-coverts. Wings and tail dusky blackish, edged with yellowish green on the outer webs of the feathers. Below dull dirty yellow, washed with olive along the sides of the body.[235]

Measurements:

Wing 2.45 to 2.54 inches, tail 1.9 to 2.2, culmen 0.47 to 0.49, tarsus 0.8.[236]

—Walter Rothschild, 1900

Geographic range: Island of Maui.

Specimens in museums: 62.

Historical information: First specimen collected 1879. Last observed 1988.

Reports and observations:

August 1…we again heard the note of the 'Akakane (*Loxops*), but it was so low that it took some minutes to locate it. At last my assistant saw one in an Ohia-tree and instantly shot it, and was lucky enough to secure the female with the second barrel.[237]

—Henry C. Palmer, 1892

Palmer discovered this species in the hills of Maui, whence he sent a fair series, though it was not very numerous. He saw it mostly creeping about among the leaves of the Ohia-trees, feeding on small insects and larvae. In its manners it was like its congener *L. coccinca* of Hawaii, but its note was lower and more plaintive. It was not at all shy. The stomachs were generally full of small beetles and other insects.[238]

That excellent observer, Mr. Perkins, describes the call of *Loxops ochracea* and the other species of the genus as a plain "*keewit*," "uttered once or repeated, and constantly to be heard. They seek," he proceeds, "their food amongst the leaves, especially at the ends of the branches, more rarely on the limbs themselves. It consists largely of caterpillars and small spiders. They also suck the nectar of the ohia flowers *(Metrosideros)*; this I saw them do but rarely, and only two of the species, *L. aurea* (sic!) and *L. cearuleirostris*. Most often, when seen amongst the blossoms, they were merely seeking insects, thereby attracted; but several times I shot specimens with the beak dripping, and on tasting the fluid found it to be, beyond doubt, the nectar of these flowers. From the other green birds, the green young and females of Loxops are readily distinguished, at any height, by their more forked tails, which, combined with their short, thick beaks, give them a very finch-like aspect." On one occasion Mr. Perkins saw "a pair of this bird building, high up in a tall ohia tree, toward the end of a branch. They came down to the ground for material, stripping off the brown down that covered the young fronds of some stunted 'pulu' ferns." On another occasion he "watched a pair sporting on the wing, now ascending, now descending, but gradually rising upwards till they became mere specks in the sky. It must have been several minutes before they finally alighted at no great distance from their starting-point. Both were splendid males."[239]

—Walter Rothschild, 1900

I found it in June to be rather rare on the slopes of Haleakala, frequenting scattered koa trees which are here not at all common. No doubt this bird is more numerous in localities where the koa abounds. The akepeuie formed little family groups, the fully fledged young accompanying their parents, now in the moult, in their search for food. The food of this, as of the other species, consists chiefly of caterpillars and very small spiders. These they obtain almost wholly from leaves of the koa, though occasionally the akepeuie is found hunting in the ohia trees.[240]

—H. W. Henshaw, 1902

Though I made a careful search I did not find this bird in February 1928, nor did I see anything of it on the survey when on Maui in July 1936. However it may still hold out in the large Maui forests.[241]

—George C. Munro, 1944

F. W. Frohawk del. et lith. West, Newman imp.

LOXOPS AUREA

Male, red form.

Munro (in Gregory 1928, 1936) searched for them in 1928 and 1936 without success, and based on the lack of reports, the species was listed as endangered by the USFWS in 1970 and by the State of Hawaii in 1982 (USFWS 1982c, 1983d, 1984d, 2006). From 1950 to 1995 there were scattered reports of Maui 'Akepa from the upper slopes of Haleakala. Over 100 individuals reported in 1961 (E 22:20) were questioned by E. Bryan (E 22:21) but remaining observations include descriptions of 1-3 individuals from experienced observers into at least the early 1980s (e.g., Richards and Baldwin 1953, Casey 1973, Scott and Sincock 1977, Conant 1981, Engilis 1990), and at least some of these could well be correct. Based on observations of 8 individuals at four locations during the HFBS in 1980 (later questioned by USFWS 2006), Scott et al. (1986) estimated a population of about 230 individuals with a patchy relict distribution, perhaps concentrated at 1100-2100 m elevation in and W of the Waikamoi watershed (see also Scott et al. 2008). Other observations in the 1970-1980s occurred near the junction of Kipahulu Valley and Haleakala Crater. Since the 1980s several reports have been by observers spending one to a few days in the field whereas intensive surveys (Baker 2001, Reynolds and Snetsinger 2001, Simon et al. 2002, USFWS 2006) have failed to confirm them, and we consider these reports as unsubstantiated. We presume that the Maui 'Akepa if it survived far into the 20th century at all, may now be extinct, or close to it, the victim of diseases and habitat deterioration resulting from cattle and pigs (Scott and Kepler 1985, Scott et al. 1985, VanderWerf 2013).[242]

—R. L. Pyle and P. Pyle, 2017

F. W. Frohawk del. et lith. West, Newman imp.

RHODOCANTHUS PALMERI

Greater Koa Finch

Rhodacanthis palmeri

Description:

Adult male. Head and throat rich reddish orange; back and upper wing-coverts dull greenish olivaceous, brightening to a dark dull orange on the back and upper tail-coverts. Wings and tail dark blackish brown, the feathers externally margined with dark dull yellow, paler and more greyish towards the tips of the primaries. Upper part of breast dull reddish orange, passing into dull orange-yellow on the abdomen,becoming much paler on the lower abdomen and under tail-coverts, which are pale yellow. Under wing-coverts and axillaries olive-greyish, washed with orange. Bill bluish grey, tip brown; feet grey with a bluish tinge, soles pale yellow.[243]

Adult female. Upper parts olive-green, washed with yellow on the forehead and brightening into yellowish green on the rump and upper tail-coverts. Wing and tail blackish-brown, the feathers externally margined with dull yellowish green. Throat and sides of body yellowish olive-green; breast and middle of the abdomen dull white, washed with green. Size similar to that of male.[244]

Young males resemble the female, but the feathers of the throat, breast, and abdomen are spotted and clouded with dusky, thus giving these parts a very different aspect. As they get older the underparts become more yellow, the throat dull orange, and the reddish orange of the head is first beginning on the forehead.[245]

Measurements:

Total length about 8¾ inches according to Palmer's measurement in the flesh, but only 7 to 7.5 in skins; wing 4.1 to 4.3; tail 2.95; culmen 0.8 to 0.87; tarsus 1 to 1.1; length of bill from gape to tip 0.75, height at base 0.6[246]

—Walter Rothschild, 1900

Geographic range: Island of Hawai'i.

Specimens in museums: 65.

Historical information: First specimen collected September 26, 1891. Last observed March 1896.

Reports and observations:

On September 28th the first specimen of the big "Finch" which I described under the name *Rhodacanthis palmeri* was shot.[247]

On October 5th Palmer writes:-" I have today been on the slopes of Mauna Loa and succeeded in getting no less than five specimens of the big Finch...[248]

This fine bird, which is by no means rare in the Kona and Hilo districts of Hawaii, was discovered by my collector Palmer, in whose honour I have named it. Many specimens were shortly afterwards collected by Mr. Perkins in the forests at an altitude of about 4000 feet. Palmer describes these birds as by no means shy, but rather fearless and easy to approach. He found them to feed on the Koa seeds, but in their stomach he also found seeds of the Alaii and other plants. In the foliage of the tall trees the females and young ones are difficult to see. The call-note is described by Palmer as a low whistle, sounding somewhat like a prolonged "week." It generally consists of two or three notes, beginning high and descending towards the end: it sounds melancholy. By imitating the cry Palmer succeeded in luring them very closely towards himself. The natives called the bird "poupou" and "Hopue," but did not seem to be well acquainted with it.[249]

—Walter Rothschild, 1900

It fed largely on the green beans of the koa. The first one I saw was hanging back downwards evidently working on a bean pod, after the manner of Pseudonestor, as described by Perkins. Old pods on the ground had a row of holes cut in them where the seeds had been. The seeds of these Hawai'i koas were large and sometimes the bird's stomachs were stuffed with large pieces which had been cut up by their sharp-edged beaks. Some had a few smaller seeds in their stomachs and one small group was feeding in the low aalii trees *(Dodonea)*. They made quite a rustling in extracting the seed from the shell surrounding it. Perkins found it occasionally feeding on caterpillars and feeding its young on them.[250]

J. G. Keulemans del. et. lith Mintern Bros imp.

TELESPIZA PALMERI, (ROTHSCH.)

(1) Male adult, (2) Female adult, (3) Juvenile

Its song or call is several whistle flute-like notes, the last ones prolonged. It seemed to flood the whole surrounding forest. We could find it only by its whistle, loud from the tops of the koas. That itself was difficult as the bird seemed to have ventriloquial powers. It required patience to secure the specimen.[251]

We collected adults and young birds, apparently still with the parent, feeding on the koa beans. Ovaries of a female in one of the groups were enlarged as if she were preparing to have another brood.[252]

—George C. Munro, 1944

Lesser Koa Finch

Rhodacanthis flaviceps

Description:

Adult male. Head and neck dull golden yellow; back and upper wing-coverts greenish olive, shading into bright olive-green on the rump and upper tail-coverts; wings and tail dark-brown, the feathers externally margined with green. Chin, throat, and upper breast dark wax-yellow; abdomen and under tail-coverts yellowish green.[253]

Adult female. Underparts of a yellowish green, much paler than in the male; above similar to the male, but the head greenish olive like the back, only a yellow wash on the forehead. The female is also a little smaller than the male, the wing measures only 3.6 to 3.7 inches.[254]

Young males closely resemble the adult females, but they are paler beneath and indistinctly spotted or clouded on the breast.

Measurements:

Total length about 7.5 inches according to Palmer's measuring in the flesh, but only 6 inches in skin; wing 3.8 to 3.85, tail 2.5, culmen 0.7 to .72, tarsus 0.95 to 1.[255]

—Walter Rothschild, 1900

Geographic range: Island of Hawai'i.

Specimens in museums: 8.

Historical information: First specimen collected September 30, 1891. Last observed October 16, 1891.

Reports and observations:

I created this genus in 1892 for the two thick-billed birds from Hawaii which I called *Rh. Palmeri* and *Rh. Flaviceps.*[256]

Palmer obtained a small series in the district of Kona, at the same places where *R. palmeri* was first collected. The smaller size and yellow head of the adult male serve to distinguish this species very easily from the much larger *R. palmeri* with its orange-red head in the adult male. Neither Wilson nor Perkins met with this bird.[257]

—Walter Rothschild, 1900

We were surprised to see that some of our "big finches" were classed as a different species. There is some doubt about the validity of this species and I shall quote from my notes to throw some light on this: September 30, 1891. In the heat of the day I shot another of the great finches. It has a little yellow over its bill. It was feeding in a koa top. Palmer killed another at the same time, a much smaller bird with golden head and neck and light yellow breast. It was also in a koa. The tree has seed pods on it and it seems they are feeding on the seeds.[258]

Had we had the slightest suspicion that there were two species I would have taken more complete notes. They might, of course, be hybrids with the ou but this seems unlikely considering the habits of the birds.[259]

—George C. Munro, 1944

J. G. Keulemans del. et. lith Mintern Bros imp.

TELESPIZA FLAVICEPS, (ROTHSCH.)

Male and female

F. W. Frohawk del. et lith.

West, Newman imp.

CHLORIDOPS KONA
Female

Kona Grosbeak

Chloridops kona

Description:

Adult female. Bill dull flesh-colour; lores dusky black. General colour above bright olive-green, passing into golden-green on the throat and belly; abdomen whitish; quill-feathers dusky black, edged outwardly with olive-green. (Sexes similar)[260]

Measurements:

Total length 5.75 inches; wing from carpal joint 3.25; tail 2; bill from gape to tip .8, height from chin to forehead .73; maxilla, width at base .52; mandible, width at base .59.[261]

—Scott B. Wilson, 1899

Geographic range: Island of Hawai'i.

Specimens in museums: 56.

Historical information: First specimen collected June 21, 1887. Last observed September 1892.

Reports and observations:

I shot a single example of this species on June 21st, 1887, when collecting at an elevation of about 5000 feet in the district of Kona on the west coast of Hawai'i, in a great tract of forest, consisting principally of koa-trees *(Acacia Koa)*; but there were also the mamane *(Sophora chrysophylla)*, the alii *(Dodoncea viscosa)*, the sandalwood *(Santalum album)*, and the bastard sandalwood *(Myoporum santalinum)*. I think that as *Loxioides bailleui*, so far as I know, feeds only on the seeds of the Sophora, it is most probable that this big Finch eats them also. During my stay of four weeks I only saw three examples of it. The specimen shot was on a tall *Myoporum*. The bird must be extremely rare, as I have since collected at almost similar elevations, where there are the same species of trees, but failed to obtain either of these Finches again, nor do the natives know them, whence I conclude that they are peculiar to the Kona district.[262]

—Scott B. Wilson, 1899

At the time this bird was found it was apparently unknown to the Hawaiians, and no native name has been learned. Endemic to the island of Hawai'i and occupying an area of about 4 miles square...[263]

It frequented the naio *(Myoporum)* trees and in breaking the hard seeds with its strong heavy beak made a cracking sound which guided us to it. Perkins never saw more than 6 or 8 in a day but Palmer on one occasion collected 12 in one day...Perkins found it a sluggish bird but I thought it active at times. When feeding it did seem sluggish, it sat still and cracked nuts but it could move about with considerable alacrity.[264]

A small squeak and a light sweet song are the only call notes recorded. A low cheep repeated at intervals. It sang more vigorously when excited, on one occasion when it lost its mate. It has not been seen since Perkins collected it in the eighteen-nineties. Henshaw failed to find it and considered our wonderful collecting area, which we thought a collector's paradise, one of the poorest over which he had hunted. It can be seen from some of Perkins' experiences how quickly the birds disappeared from some localities.[265]

There is a remote possibility that it will survive on some of the rough lava flows at a high elevation.[266]

—George C. Munro, 1944

J. G. Keulemans del. et. lith Mintern Bros imp.

CHLORIDOPS KONA, (WILS.)

(1) Adult, (2) Juvenile

Kona Finch, *Chloridops kona,* feeding chicks with looper caterpillars. (JULIAN P. HUME)

F. W. Frohawk del. et lith.

West, Newman imp.

PSITTIROSTRA PSITTACEA

ʻŌʻū

Psittirostra psittacea

Description:

Adult male. Head and neck gamboge-yellow, all the rest of the upper parts olive-green inclining to yellow on the rump; whole of the under surface greenish yellow with the exception of the breast, which is grey; remiges and rectrices dusky brown margined with olive-green; irides dark hazel; bill and feet pinkish.[267]

Adult female. Head and neck olive-green above and grey beneath; the rest of the under surface greyish white; under tail-coverts pure white.[268]

Measurements:

Adult. Total length 6.30 inches, wing 3.85, culmen .70, tarsus .95, tail 2.20.[269]

—Scott B. Wilson, 1899

Geographic range: Islands of Kauaʻi, Oʻahu, Molokaʻi, Lānaʻi, Maui, and Hawaiʻi.

Historical information: First specimen collected 1779. Last observed 1989.

Specimens in museums: 323.

Reports and observations:

A Bird with a yellow head, which from the structure of its beak, we called a Parroquet, is likewise very common;[270]

—Captain Charles Clerke, 1779

This well-known species is distributed throughout the Sandwich group, and I obtained specimens from every island save Oahu and Maui, on the former of which I have good reason to believe it has become extinct or else extremely scarce. I cannot detect any appreciable difference between examples from the various islands, although I think those I obtained on Lanai are brightest in plumage. The size and shape of the curiously formed bill varies considerably, especially in the males: … Next to *Vestiaria coccinea,* it is perhaps the most noticeable of the forest-birds of the islands, the bright yellow head and neck of the adult males rendering them very conspicuous in their straight dashing flight from tree to tree. The immature males and females, which lack this distinctive feature, might easily be mistaken for the sombre-coloured *Phaeornis obscura*; but the constant twittering which the Ou almost invariably makes while feeding at once betrays its identity. Freshly killed examples possess a peculiar scent, which I did not observe in any other forest-dwelling species; it is probably due to their extremely varied fruit-diet.[271]

I think that I found it most plentiful among the trees which clothe the abrupt sides of the deep ravine running down to the leper settlement on the island of Molokai; and very lovely these little birds looked, flying continually to and fro, up and down this stupendous gorge—their yellow necks flashing in the bright sunlight, as they darted out from among the dark green ohias or from the silvery foliage of the kukui *(Aleurites triloba).* The food of Psittacirostra consists entirely of fruits, and chiefly of that of the ieie *(Freycinetia arborea)*, the ripe seeds of which I found in most cases in the stomach when dissecting specimens; I noticed also, particularly in one place on the outskirts of a forest in the district of Kona, that a very large proportion of the fruits of the climber were eaten away at the apex, and here I shot a good number of examples. I killed others as they were busily engaged in feeding on the small crimson fruit of the wild mulberry *(Morus papyrifera)*, the juice of which had dyed their throats a deep crimson.[272]

Necklaces, "leis," used sometimes to be made from the bright green plumage of the back and underparts of this bird, but they were commonly used in combination with the black feathers of *Acrulocercus nobilis* and the scarlet feathers of *Vestiaria coccinea*. I saw a wreath thus made at Olaa in the district of Puna, which I attempted to purchase, but the native woman wanted a higher price than I was inclined to give.[273]

—Scott B. Wilson, 1899

Like the Apapane, I have observed the Ou making long-sustained flights from the palis of the large valleys, that carried them readily from one valley to another. At such times they rarely, if ever, soar or circle about, but set off directly for the fresher fields with a show of knowledge and determination that makes them while on the wing, easily distinguished from their neighbors, as far as they can be seen.[274]

—William A. Bryan, 1908

On Kaua'i, Stejneger (1887) considered the 'O'u "rather rare" but Bryan and Seale (1901, 1915) found them common at low elevations in the 1890s. By the 1940- 1970s observations were primarily restricted to singles or pairs in the Alaka'i Swamp (e.g., Richardson and Bowles 1964; Berger 1972, 1981; Banko 1986; Conant et al. 1998, Gorresen et al. 2009; E 2:52, 22:3, 24:22-23, 28:102, 32:51-54, 37:29; HRBP 6707), although 6 together and at least 10 over all were observed at the head of the Koaie Stream 1 Jun 1968 (E 29:19). The 'O'u was listed as endangered by the USFWS in 1967 and by the State of Hawaii in 1982 (USFWS 1982, 1983, 1983b, 2006), with exposure to avian malaria considered the primary (but not the only) threat (Atkinson 1977; Scott and Kepler 1985; Scott et al. 1985, 1986). Sincock et al. (USFWS 1983c) estimated a total population of 62 in 1968-1973 and Scott et al. (1986) estimated only 10 remaining in 1981, based on a sighting of only one individual on transects during the HFBS; similarly, only one could be found during a survey in 1984 (Scott and Kepler 1985). The last confirmed observation was of one on 17 Feb 1989 (AB 43:370, HWN 4[2]:1; cf. E 53:49) and the 'O'u is now presumed extirpated from Kaua'i (Reynolds and Snetsinger 2001, Forster et al. 2004, USFWS 2006, Scott et al. 2008, Paxton et al. 2016; see also below). We consider additional reports of 'O'us on Kaua'i in 1991-2010 as unsubstantiated.[275]

On O'ahu, up to 46 specimens were collected, beginning with one by Bloxam in 1825 (Olson 1996a) and concluding with one by Mann in 1869 (Rothschild 1900, Banko 1979). Subsequently the 'O'u was not found commonly, undoubtedly because it had already declined severely there by the time more collectors got there in the 1880s-1890s (Perkins 1903, Munro 1944). Wilson could not find them in 1887-1888 (Banko 1986) and believed them "extinct or extremely scarce" on O'ahu at that time (Wilson and Evans 1899). A pair observed by Perkins (1903) in 1893 at about 914 m elevation on Mt. Ka'ala is the last confirmed observation from O'ahu; Perkins (in Evenhuis 2007:111) considered them possible transients from Kaua'i or Moloka'i. Bryan (1901a) possibly observed one in Moanalua Valley in Oct 1899.[276]

On Moloka'i 'O'u were common in the late 1800s along the cliffs above Kalaupapa (Wilson 1890a, Wilson and Evans 1899) and in the interior (Schauinsland 1906; see E 57:76-79). Bryan (1908) found them widely distributed but uncommon in Apr-Jun 1907 and collected 16 individuals, but these were the last substantiated record for the island (Munro 1944; cf. E 27:90-91). Perkins (1903) and others considered 'O'u "excessively common" on Lana'i during the 1890s (Munro 1944), and were still considered common there in 1913 (Munro 2007) but they quickly vanished, 5-6 specimens collected in 1923-1927 (Banko 1986) and observations in 1931 ("rarely") by Munro (in Gregory 1932) being the last reports (cf. E 6:3, Hirai 1978; see also below). On Maui, Finsch (1880) noted 'O'u as abundant above Olinda in 1879. Fifteen specimens were secured by early naturalists in 1892 and 1901, the last by Henshaw on 21 Jun 1901 (Banko 1979, 1986; see also below). Banko (1986) credits Baldwin with a possible observation on Maui in 1945 but it was not mentioned by Richards and Baldwin (1953) and we assume this report is an error.[277]

On Hawai'i, 'O'us were brought alive to Captain Cook (Cook and King 1784; misidentified as Hawaii Creeper and 'Elepaio by Beaglehole 1967:630) and they were noted by late-19th-century naturalists as being common or abundant and widespread (Perkins 1893, 1903; Rothschild 1900; Henshaw 1902). They were generally found between 300 and 1200 m elevation but were collected as low as Hilo in 1875 (Sclater 1881) and were reported by Palmer up to 2,100 m (Rothschild 1900). Subsequently, none were reported for over 30 years (Banko 1986) and by the 1930-1950s only small numbers were detected in the upper Ola'a Forest Preserve and in the Puna District, at elevations of 750-1,220 m elevation (e.g., Baldwin 1944, 1953; Richards and Baldwin 1953; E 5:55, 11:64, 16:41). Scattered observations of 'O'u in these and other areas of se. Hawai'i continued through the 1970s (e.g., Dunmire 1961; Conant 1980b, 1981; USFWS 1982, 2006; Gorresen et al. 2009; E 20:89, 35:21, 40:158). Only 33 individuals were recorded on transects during the HFBS in 1977-1979 (32 in Hamakua and one in Puna) and, based on this, Scott et al. (1986) estimated about 400 individuals remaining at 900-1,500 m elevations at the time. Highest densities were in the upper Waiakea Forest Reserve, in forests that have since been destroyed by volcanic eruptions (Snetsinger et al. 1998, Reynolds and Snetsinger 2001). The last substantiated observation was of 2 individuals in the Ola'a tract 14 Jun 1986 (cf. E 47:65). Considering the plummeting population trajectories on other islands and the preference of this species for lower elevations, we now presume the 'O'u is extinct on Hawai'i and all other Hawaiian Islands (Reynolds and Snetsinger 2001, Scott et al. 2008, Gorresen et al. 2009). Based on Poisson analyses of persistence probabilities using confirmed and unconfirmed records, Elphick et al. (2009) estimated that the 'O'u was extirpated on Hawai'i in 1984-1990, with upper limits of 1998-2001. Values for estimated extirpation years on other islands include 1984-1993 (upper limit 2002-2007) on Kaua'i; 1911 (1928) on Moloka'i; 1932-1934 (1948-1959) on Lana'i; and 1905 (1919) on Maui.[278]

—R. L. Pyle and P. Pyle, 2017

ʻŌʻū collected by Scott B. Wilson on Kauaʻi, March 29, 1891.
(COURTESY OF NATURALIS BIODIVERSITY CENTER LEIDEN, NETHERLANDS)

F. W. Frohawk del. et lith.

West, Newman imp.

DREPANIS PACIFICA

Mamo

Drepanis pacifica

Description:

Glossy black, with the exception of the lower part of the body, the rump, the tail-coverts both above and below, the feathers of the tibia and those of the anterior margin of the wing, which are of a fine crocus-yellow; the larger primary wing-coverts and under wing-coverts white, the former mottled with blackish grey, and the latter tinged with yellow. Remiges brownish black, tipped with dull white on the external vane of the five outer primaries, and both vanes of the next four, as well as those proceeding from the olecranon. Four middle rectrices glossy black, the rest more or less brown, and showing a patch of dull white near the tip, which though indistinct on the inner feathers becomes very distinct on the extreme pair. Bill and legs apparently deep brown.[279]

Measurements:

Total length about 8 inches, wing from the bend 4.12, tail 2.5, bill from forehead along the culmen 1.75, from gape in a straight line 1.5, tarsus 1.31, middle toe without claw .56, hind toe .43.[280]

—Scott B. Wilson, 1899

Geographic range: Island of Hawai'i.

Specimens in museums: 11.

Historical information: First specimen collected 1779.
Last observed 1898.

Reports and observations:

... one of which is something larger than a Bullfinch; its colour is a fine glossy black, the rump, vent, and thighs deep yellow (it is called by the natives *hoohoo*).[281]

—Captain Charles Clerke, 1779

The feathers of birds were the most valued possessions of ancient Hawaiians. The feathers of the mamo were more choice then those of the *oo* because of their superior magnificence when wrought into cloaks (ahu).[282]

—David Malo, 1838

Of this extremely rare and apparently extinct species I obtained two specimens from a collection which was formed by the late Mr. Mills of Hilo in Hawaii, some thirty years or more ago. The fact of its native name "Mamo" being the same as that used for the war-cloaks mentioned below seems to imply that they received it from this bird, and that they were originally chiefly wrought of the beautiful golden yellow feathers from its back and vent, which are much deeper in colour, as they are larger and longer, than the axillary tufts of *Acrulocercus nobilis*.[283]

—Scott B. Wilson, 1899

I am glad to be able to state that the "Mamo" is not yet quite extinct, because Palmer sent me a skin with the body in spirits.[284]

April 16th (1892).- Broke the camp up at 6 A.M. and pushed on till 4 P.M. The old bird-catcher Ahulan was leading to cut the trail, whilst Holi and myself came next, followed by the others, who were a long way behind. We had not gone more than three miles, when I heard a call from the other side of the gulch, and thought it was a native calling, but immediately afterwards a bird flew across, and I saw in a moment it was the bird we were after. I was going to follow it up to shoot it, but Ahulan begged me not to shoot it as it would scare the other away, which I had heard calling a little way off. Ahulan fixed the snares and bird-lime on a haha, which growed out on a tree-fern, and which has flowers somewhat like those of a fuchsia. Ahulan fulfilled his promise and caught the Mamo! He is a beauty, and takes sugar and water eagerly and roosts on a stick in the tent. I now feel as proud as if someone has sent me two bottles of whisky up.[285]

—Ted Wolstenholme (in Rothchild), 1900

In spite of two or three days' more work, no second specimen was caught or even noticed, and at last the natives declared that Palmer's "assistant" was mistaken about having seen two, and that there was no other specimen about there. When their food was finished and they had no hope of getting any more "Mamos," the whole party descended to where Palmer stayed. On 21st Palmer killed and skinned the Mamo.[286]

—Walter Rothschild, 1900

In April 1892, Henry C. Palmer was laid up after a horse accident, allowing his assistant Ted Wolstenholme to search for Mamo with the bird-catcher Ahulan. After days in the field, Wolstenholme triumphantly returned to camp with a live bird, and an unknown hand took this unique photograph. So prized was his possession that Wolstenholme tied the bird's ankle to his waistcoat with string to prevent escape. Unfortunately for the bird, Palmer, who was embittered and without sentiment, promptly killed and skinned it. The lower bird depicted on page 198 is the above-mentioned individual.

...Mr. H. W. Henshaw, the distinguished American ornithologist, who now resides in Hawai'i, writes to me (dated October 9th, 1899) as follows:-

Drepanis pacifica is still a living species, though unquestionably very rare. No doubt on the verge of speedy extinction. About a year ago last July I found what, no doubt, was a family of Mamos in the woods above Kaumana. I am sure that I saw at least three individuals, possibly four or five. They were flitting silently from the top of one tall ohia-tree to another, apparently feeding upon insects. The locality was a thick tangle, and a momentary glimpse of a slim, trim body as it threaded its way through the leafy treetops was all I could obtain. After about two or three hours I succeeded in getting a shot at one bird in the very top of a tall Ohia-tree. It was desperately wounded, and clung for a time to the branch, head downwards, when I saw the rich yellow rumpmost plainly. Finally, it fell six or eight feet, recovered itself, flew round to the other side of the tree, where it was joined by a second bird, perhaps a parent or its mate, and in a moment was lost from view. I need not speak of my disappointment, which was bitter enough, for I had looked upon that bird as absolutely mine own. Of the others I saw no more, though I have repeatedly visited the locality again.[287]

—Letter from H. W. Henshaw, 1900

J. G. Keulemans del. et. lith Mintern Bros imp.

DREPANIS PACIFICA, (GM.)

There is every reason to believe that this famous bird is extinct, or so nearly so as to leave little hope that it will ever again be seen by human eyes. Nevertheless the writer saw at least a pair, possibly a whole family, in the woods of Kaumana, July, 1898. A year later I was assured by a native who had just come down from the deep forest, not far from the above locality, that a few days previously he had heard the note of a mamo near by but did not see it. His efforts to secure the bird, if he found any as he promised, were not successful. This fact brings the living history of the bird down to the year 1899.[288]

We know next to nothing of the habits of the mamo. The birds I saw in Kaumana were very active, and evidently were in pursuit of insects which they were hunting in the very tops of the tall ohias. The birds flight was not rapid, but was smooth and well sustained, and the bird on the wing reminded me of nothing so much as the cuckoo. Though I observed the birds at intervals for more than two hours, I did not hear a single note.[289]

—H. W. Henshaw, 1902

The Mamo is now a very rare bird and it is quite possible that any year may be the last of its existence. Formerly it was without doubt of wide range over the island of Hawaii since it is known to have occurred both in the leeward and windward forests as well as in the Kohala Mountains. Unlike the Oo, which after the yellow axillary feathers were plucked out, could be, and sometimes was, liberated practically uninjured, the Mamo entirely denuded of its yellow feathers would have been in a sad plight, and would almost certainly have succumbed to such rough usage. It is quite certain that up to about 20 years ago these birds still existed in some numbers in the forests of Hilo district, for at the time of the great lava-flow of 1880 a considerable number were shot for the sake of the yellow feathers, as many as twelve having been obtained by a native bird-hunter in a single day. None of these were preserved as entire skins, the yellow feathers alone being saved.[290]

That the Mamo is now extremely rare is certain, for Mr. W. H. Henshaw, the expert American ornithologist, has not up to the present time obtained a single example, although he has resided only a few miles from a noted and perhaps now the only locality for this bird, whence was obtained the only specimen procured of late years.[291]

Very little is known as to the habits of the Mamo, and the accounts that have been handed down, and the information one can glean from the few existing natives who have seen it, do not always agree. Some speak of it as very tame, others as wild and shy, but the latter account seems generally to refer to localities where, and to a time when, it was systematically hunted. All agree as to its fondness for the nectar of the arborescent lobelias, and it was at the blossoms of these that it was usually captured.[292]

—R. C. L. Perkins, 1903

F. W. Frohawk del. et lith. West, Newman imp.

DREPANIS FUNEREA

Black Mamo

Drepanis funerea

Description:

Adult male. Entirely dull black above and below, the feathers blackish grey at base. Primaries sooty black, outer webs of a huffy or dirty white, passing into blackish at base. Feet black, soles yellowish (in skin). Bill black, nasal operculum and base of upper bill yellowish (in skin). This colour is undoubtedly quite natural and must have been brighter in life.[293]

Female similar to the male, but smaller, bill shorter.[294]

Measurements:

Total length about 8 inches; culmen 2.0; bill from base to tip of upper mandible 2, under mandible from base to tip 1.7, thus being .3 shorter than the upper; height of bill at nostrils 0.3; wing nearly 3.9 inches (damaged at tip), tail 2.8, tarsus 1.15.[295]

—Walter Rothschild, 1900

Geographic range: Islands of Moloka‘i and Maui.

Specimens in museums: 10.

Historical information: First specimen collected June 18, 1893. Last observed June 1907.

Reports and observations:

The Oo-nuku-umu is one of the rarer island birds and is now confined to the higher parts of the forest on Molokai, where the ground is soft and boggy. At the time when I discovered it these woods were in absolutely natural condition, but since that time both cattle and deer have run through them and they are in most parts much less dense and less wet than they were in 1893. The bird is essentially one of the underbrush, and I never saw it alight high up in a large tree, and only very rarely was it seen at more than twelve feet from the ground. It is a true honey-sucker, and not one of those obtained by me contained animal food.[296]

With that pugnacity which is said to be equally characteristic of the Mamo, it continually drove the smaller red birds from the bush, to be itself in turn driven off by the superior strength of the Oo, but always returning after a few moments to the same flowers.[297]

More usually they readily approach and even follow the collector, or alight on a branch at a distance of only a yard or two, moving occasionally from place to place and turning the head this way and that to make a thorough scrutiny. For this reason they are sometimes even difficult to shoot, as they keep following the collector as he retires to a suitable distance for a shot, and when they fly off at last it is for good and all.[298]

They readily respond to an imitation of their call note, which much resembles that of the Mamo, but sometimes they utter a loud cry of extraordinary clearness, repeated at short intervals, such as I have never heard imitated by the old natives, who well knew the call of the latter. No doubt this loud cry takes the place of a true song and differs rather in degree than kind from the more gentle call note. The Oo-nuku-umu would appear to be the rarest of all the birds of which I obtained specimens. On one occasion when not prepared to collect birds I saw seven individuals in a single day, but this was quite exceptional, and when wanting a specimen I have spent a whole day or several days in its haunts, and in exclusive search for it, without seeing even a single one, and this at a time when I had become perfectly familiar with its habits.[299]

—R. C. L. Perkins, 1903

The severe labours undergone by Mr.Perkins, who has for some months past been exploring Molokai, one of the Sandwich Islands, have met with a suitable reward in the unexpected discovery of a new species of Drepanis, of which he has sent home several specimens; and, in describing it at his request, I feel bound to express the satisfaction of the Joint Committee under whose direction he has been working, coupled with my own regret that his modesty hinders him from introducing his discovery to the world. Its sombre plumage and the sad fate that too probably awaits the species induce me to propose for it the name of DREPANIS FUNERA.[300]

—Alfred Newton, 1893

The Hoa, or "Oo-nuku-umu" as Perkins calls it, is exceedingly rare, so rare, indeed, that seven weeks of continuous field work in the forests of Molokai, its only habitat, were necessary before seeing or hearing, let alone securing a single specimen.[301]

...Mr. Theodore Meyer, a resident of Molokai, secured six specimens from the same locality, in six months of more or less continuous collecting. He believed the species to be extinct, as neither he nor any of his friends from then until now had been able to find the bird in any locality.[302]

We were working in a particularly dense and unpromising jungle of ieie, when I heard a clear, gentle, even, inquiringly whistled "H-o-a" called by some bird not more than fifty yards away. It was a moment of intense excitement. There was little doubt in my mind but that note was that of the bird I had so long sought. Making signs to the native to hurry to me with the gun, I began to imitate the whistle call as best I could. To my delight the curious black bird with the wonderfully curved bill, the object of my seeking, flew down and lit in the tree within ten feet of me! It was evident that curiosity aroused by my chopping had brought the bird to the spot to investigate.[303]

Actively hopping and flitting from limb to limb, scarcely stopping a second, eyeing me sharply all the while, it persisted in keeping so close to me that I could not with safety fire even my auxiliary barrel. Yet the restlessness of the bird made it apparent that I must fire soon or lose my chance, for if the bird got fifty feet from me in the tangle of trees and vines on the side of the ridge, it would, in all probability be gone from sight and forever. It was after all little wonder that gun and ammunition failed me, for both had been soaking wet for three days. The next few seconds were moments of painful indecision. To break and reload my gun would surely frighten the bird; to use the heavy charge in the left barrel was almost criminal, but my only recourse. Without further delay I availed myself of the first opportunity when the bird was screened by limbs and fired. The feathers flew: my first Black Mamo had been shot.[304]

—William A. Bryan, 1908

Perkins collected a series of skins, and private collectors later depleted the district above Pelekunu valley where Perkins had collected. Neither Alanson Bryan or I could find any there in 1907, but Bryan secured 3 male specimens at Moanui farther round the island to the east. The survey in 1936 found only a rumor that one had been seen in the Wailau valley a few years before. Donaghho traversed much of the Molokai forest in 1937 and found no trace of it.[305]

—George C. Munro, 1944

Laysan Honeycreeper

Himatione fraithii

Description:

Adult Male. Head, throat, breast, and upper abdomen of a bright scarlet-vermillion. With a faint tinge of golden orange, which is more pronounced in freshly moulted and living birds; lower back and upper tail-coverts similar in colour to the head, rest of upper parts orange-scarlet; the feathers of these parts are grayish brown at base, with a pale shaft-stripe and broadly tipped with orange-scarlet, those of the head and back of neck with distinct whitish spots before the red tips; on the hind neck all the basal part is quite concealed and only the red tips visible, but on the back of the neck the sub-terminal colour becomes somewhat conspicuous; lower abdomen dull ashy brown; quills dusky brown, with blackish shafts, too, are pale; primaries externally narrowly edged with whitish, inside with buff; secondaries with pale scarlet outer edges. Bill and feet black; iris reddish yellow.[306]

Adult female. Similar to the male, but the red somewhat paler in tinge.[307]

The present species resembles Himatione sanguinea, from the Sandwich Islands, but is easily distinguished by its shorter bill and its scarlet-vermillion colour instead of the blood-red darker colour of *Himatione sanguinea.*[308]

J. G. Keulemans del. et. lith Mintern Bros imp.
HIMATIONE FRAITHI (NEWTON).

Measurements:

Total length about 5½ inches, wing 2.55 to 2.7 (average 2.65), tail 2.4, culmen 0.55, tarsus 0.9.[309]

—Walter Rothschild, 1900

Geographic range: Island of Laysan

Historical information: First observed 1828. Last observed 1923.

Specimens in museums: 105

Reports and observations:

Palmer says of this bird:-"This is by far the rarest of the Laysan-Island birds, though I have observed a fair number, generally in pairs. They are very active in their movements, flitting about in the scrub. It feeds on very small insects as a rule, but I have also observed it sucking honey from flowers. It has a low sweet song, consisting of several notes, but very seldom gives utterance to any sound, except during the breeding season. Mr. Freeth tells me it is in full song in January and February, when the plumage also has a beautiful golden gloss over the red. So no doubt the breeding season lays between that time and June, when I met with full-grown young ones."[310]

June 17 … A most touching thing occurred: I caught a little red Honey-eater (*Himatione freethi*.-W.R.) in the net, and when I took it out the little thing began to sing in my hand. I answered it with a whistle, which it returned and continued to do so for some minutes, not being the least frightened.[311]

—Henry C. Palmer, 1891

The Laysan Honey-eater is a brilliant little bird, about the size of a warbler, and very attractive when seen flitting here and therein the soft green of chenopodium bushes. Its plumage is of a lustrous scarlet vermilion, brightest on the crown with wings, tail, and abdomen a dull sepia.[312]

They are most abundant in the interior of the island near the open plain bordering the lagoon. Here on the extensive beds of succulent portulaca they may be seen throughout he day, busily walking about like pipits, either gathering insects or drinking honey from the numerous half-blown buds. The brush-like tongue of these creatures renders the gathering of honey, and such tiny insects as may infest the interior of corollas, an easy task. In fact it was no uncommon occurrence to see one go from flower to flower, and insert its bill between the petals of a nearly opened bud, with a certain precision and rapidity which suggested a hummingbird, except of course that the *Himatione* was on its feet.[313]

—Walter K. Fisher, 1902

The honey eater is not common on Laysan. There were four birds of this species that roosted on an old rope which hung across the corner in one of the sheds. Here each night we would see them huddled closely together. An occasional bird would flit into our workroom in quest of millers. They were to be seen about the island in the tall tussocks of grass, where we found a few nests, all of which contained young birds or well incubated eggs. Among the old specimens left by the feather hunters we found several skins of the honey eater. There are possibly 300 living birds of this species.[314]

—Homer R. Dill, 1911

April 8. As far as I can see with the glasses and from our hurried trip down the island there is not a living bush or twig or spear of grass left on the whole island outside of the two poor cocoanut trees and the 3 bushes near the house that Sula piscator has preempted—and two of the latter are leafless! Apparently the swarming host of rabbits has gone the limit. The once green island is today a desolate sand waste that might as well be a salt sink in one of our barest sand deserts.[315]

April 15. The only real, bright spot was when we came on two of the almost extinct Laysan Honey Eater (*Himatione, fraithii*), a charming little red mite of a thing that has well-nigh passed with the passing of the vegetation. This is the same place Reno saw his 3, so as far as we now know we were looking at 2 of the 3 last survivors. They had done their best to adapt themselves to the changed environment and were scrambling about among the rocks and on the guano earth picking up small flies. Their charming song is out of proportion to their size. Altogether they cheered my morning immensely, although their future can be only extinction unless a more rapid restoration of the island can be effectual than I think possible.[316]

April 18. As I finished, I heard a weak but charming song behind me and whirled to find one of our pair of Laysan Honey Eaters singing his heart out for me.[317]

April 23. Rain and mist—driving N. gale—clearing to sandstorm in afternoon. Hell of a day.

And now I must pay for my fun! Awoke to the shattering tune of the tent straining at every peg before a driving north wind and soaking mist… This afternoon the clouds broke at times, but the gale increased until the whole island seemed to be walking—a regular cursed sandstorm of the worst sort.[318]

April 24. Cloudy. North gale—sandstorm. Hell of a day. The gale and blinding sandstorm continues unabated and life in the open or in the tents or tumble-clown shacks is equally unbearable.[319]

One of the last three individuals of the Laysan 'Apapane perched on a piece of coral rock.

April 25. Broken clouds—NNE gale—sandstorm (third day) hell on earth. The terrific cold gale and sandstorm continues unabated—almost unbearable—getting on our nerves—third day however, wind a trifle E of N and scudding clouds broken with moments of sun, so we pray to heaven it may go down with the sun tonight.[320]

April 26. N gale unabated—rain squalls—less sand blowing—hell continued. Fourth day and night show no abatement of the N gale and driving sandstorm.[321]

—Donald R. Dickey, 1923

Donald Dickey left Laysan on April 30, 1923. His journal contains no information about the Laysan Honeycreeper after April 18. Alexander Wetmore, the leader of the 1923 Tanager Expedition, returned to Laysan Island on May 13, several weeks after the devastating four-day sandstorm. He wrote the following summary for the Laysan Honeycreeper:

April 8-30, 1923. Three seen at intervals about piles of rock at the southern end of the lagoon. May 13. These birds have not been seen since the heavy storm after the twentieth of April and it is supposed that they have perished. I have visited their haunt on a number of occasions without finding them.[322]

—Alexander Wetmore, 1923

Video of one of the last Laysan Honeycreepers can be seen at this link: https://www.youtube.com/watch?v=WS_ob9_jCH4

Po'o-uli

Melamprosops phaeosoma

Description:

The Sexes are similar with adults being brown above and pale gray below. They have a black mask which extends from the forehead to the throat and chin to a point behind the eye. The crown is gray and the flanks and under-tail coverts are cinnamon. The bill is short and black. The legs are long and pale. Immatures are drab brown above and buffy below and have a less extensive black mask. The bill is dark with a white lower mandible which apparently darkens from the base as birds get older. The legs are dark brown.[323]

The last surviving Po'o-uli, thought to be at least 8 years old, died of old age November 26, 2004. It is pictured here at the Maui Bird Conservation Center shortly before his death. (ZOOLOGICAL SOCIETY OF SAN DIEGO)

Measurements:

Measurements (mm) for holotype and paratype, respectively: total length, 135 and 122; unflattened wing-chord, 74 and 72; tail, 38 and 36.5; exposed culmen, 15 and 14.5; tarsus, 27 and 25.[324]

Geographic range: Island of Maui.

Historical information:

First observed 1973. Last observed 2004.

Specimens in museums: 3.

Skin study of the Po'o-uli, *Melamprosops phaeosoma.* Specimen is in New York. (JULIAN P. HUME)

Reports and observations:

A husky Hawaiian honeycreeper drably colored in shades of brown, the Po'ouli (pronounced poh-oh-U-lee and translated "black-faced") was named for its most striking feature by M. Kawena Pukui, noted authority on Hawaiian language and culture. This naming followed the bird's remarkable discovery in 1973 by T. Casey (TC), J. Jacobi, and other college students on a University of Hawaii expedition to the rain forest wilderness of Haleakala Volcano on Maui Island.[325]

The Po'ouli seldom calls or sings and typically is seen long before it is heard. It is encountered singly, in pairs, or in family groups, often in association with other foraging insectivorous honeycreepers. A Po'ouli typically hops into view along a tree limb, head down, methodically searching bark and epiphytes for food. It pauses only to rip apart loose bark, lichen clumps, or moss cushions for hidden invertebrate prey. Its diet is yet another strange example of specialization among the Hawaiian honeycreepers, for besides consuming arthropods and fruit, the Po'ouli eats extraordinary numbers of land snails.[326]

The po'ouli population was estimated at 140 ± 280 birds in the early 1980s (Scott et al., 1986), but estimates of population size and density are not accurate and are considered imprecise due to the species' low density and cryptic behavior. Po'ouli numbers and range have since declined to a tiny population over at most a few hundred hectares.[327]

Thorough surveys of the historical range in 1997 to 2000 located only three birds, all in Hanawl Natural Area Reserve, and no others have been located since these birds were color-banded in 1996 and 1997 (Hawaii Department of Land and Natural Resources, unpubl. data).[328]

Following an attempted translocation in 2002 to form a breeding pair and efforts to capture the remaining Po'ouli for captive breeding, one of the three known remaining birds died of natural causes in captivity in 2004, and the other two have not been seen since 2003 and 2004. Currently it is not known whether any po'ouli remain in the wild.[329]

An extremely rare Hawaiian forest bird that was hoped to be part of a last ditch effort to save the species has died. The poouli, a male that had been living at the Maui Bird Conservation Center in Olinda, Hawaii since September, passed away close to midnight on Friday, November 26th, 2004.[330]

F. W. Frohawk del. et lith. West, Newman imp.

CIRIDOPS ANNA

‘Ula-‘ai-hawane

Ciridops anna

Description:

Adult. Lores and forehead velvety black, this colour gradually shading into the ashy grey of the crown, nape, and hind-neck, which colour again shades off into the dark sepia-brown colour of the back. Rump and upper tail-coverts dark glossy red. Tail-feathers uniform black. Primaries and secondaries black, only the outer webs of the last three secondaries earthy brownish buff (nearest to Ridgway's "clay-colour" on plate v. fig. 8); scapulars and tips of some of the greater wing-coverts of the same colour. Feathers on the sides of the head and neck, chin, and throat black with silvery-grey shaft-stripes. Breast down to the middle of the uppermost part of the abdomen black. Middle of abdomen, vent, and under tail-coverts tawny brown. Sides of abdomen largely glossy red. Under wing-coverts blackish.[331]

Measurements:

Wing 3 inches, tail 1.75, culmen 0.43, height of bill at base 0.2, tarsus 0.8, middle toe with claw 0.76, hind toe with claw 0.64.[332]

—Walter Rothschild, 1900

Geographic range: Island of Hawai‘i and possibly Moloka‘i.

Specimens in museums: 5.

Historical information: First specimen collected ca. 1859. Last observed 1892.

Reports and observations:

This is a bird of remarkable beauty, its peculiar combination of colors producing a most harmonious and elegant effect.[333]

—Sanford B. Dole, 1879

I procured a stuffed specimen from the Hon. C. R. Bishop, which had been obtained by the late Mr. Mills of Hilo. Mr. Bishop has a very much finer example remaining, with more grey about the head and neck, taken by the same gentleman. I used to hear repeatedly of the "Ulaaihawane," by which name it is well known to the natives, who told me that it feeds on the fruit of the Hawane palm, whence its name—Ula (red), ai (to eat), Hawane (the

Pair (female above, male below) of 'Ula-'ai-hawane, *Ciridops anna*, feeding in the crown of a *Pritchardia* palm with another pair in the background. (JULIAN P. HUME)

> Hawane palm); and therefore I have little doubt that it will be found, perhaps in some numbers, in the upland region of the interior, which I was unable to explore. The present specimen—now in the collection of the Hon. Walter Rothschild—has not the sex marked; so it is impossible to say whether the male differs from the female or not. My friend Mr. Francis Spencer, writing to me quite recently, says that his natives had seen the bird in the swampy forest-region above Ookala on Hawaii, and his description leaves no doubt of its identity.[334]
>
> —Scott B. Wilson, 1899

This beautiful and finch-like bird was first described by Mr. Dole from specimens taken on the island of Hawaii where alone the species has been found. These mounted birds, two in number, were in the collection of the late Mr. Mills and presumably, were taken by natives not far from Hilo, perhaps in the Olaa district. There is no evidence, however, of the exact locality whence they were derived, and it is possible that they may have come from some other part of the island.[335]

No collector has met with the species recently, save a single individual shot for Mr. Palmer by a native on Mt. Kohala in February, 1892. From the fact that the principal food of the bird (as the natives say and as its name implies) was derived from the hawane palm, there would seem to be a reasonable hope that the bird may exist in the more elevated districts of the interior where alone this palm grows in abundance. However, even if the species is still extant is must be in very small numbers. To all intents and purposes the hawane finch may be looked upon as extinct, and as furnishing another melancholy example of the mysterious fate which has overtaken so many Hawaiian birds.[336]

—H. W. Henshaw, 1902

This peculiar bird is undoubtedly one of the rarest in the world. There is no other specimen of it in Europe, and only one is known to be in the Bishop Museum in Honolulu, this making a total of only three specimens known in collections.[337]

It was of course Palmer's principal tasks to rediscover *Ciridops anna*, and he succeeded in getting one specimen, which was shot by a native on the 20th of February (1892) near the head-waters of Awini on Mt. Kohala, Hawaii. There, the natives sent him a message, the "Ulaaiwhane," as Palmer says it is properly spelt, was still seen in numbers, but when Palmer arrived on the spot the former report was denied and they said the bird shot was the only one seen there! The "Loulu" palms, they said were the trees formerly frequented by the "Ulaaihawane," and that they were scarce now, although in some deep gulches there might still be some more." Palmer, after staying a short time at the place and seeing no sign of the bird, was apparently much disappointed with the place, and he concludes with the following words:- "Undoubtedly there are more 'Ulaaihawane' in the place where the one was shot, but if a bird is rare in such a country one might travel about for months and not find one. I therefore thought it wiser to offer the natives a fair price for specimens and to go personally to a more promising place. I have shown the natives at the place how to make a skin, and they have promised to do their best.[338]

Palmer's hopes, however, were not realized, for no further specimen has turned up as yet, and I am afraid this bird is nearly extinct.

—Walter Rothschild, 1900

Conclusion

Can We Save the Few That Remain?

The first people arrived in the Hawaiian Islands approximately 1,000 years ago. Ever since, the remote archipelago's biological treasures have been under continuous attack. Species which took millions of years to evolve have been decimated in a geological blink of an eye. Today, only scattered remnants of native ecosystems and dwindling endemic bird populations survive. Vast areas of original dry and mid-elevation forests have been converted to agricultural and urban land use. Livestock and introduced game animals have severely degraded mid- to high-elevation forests. To solve the widespread damage caused by free roaming cattle, sheep, goats, horses, deer, and pigs, millions of alien trees were planted in Hawai'i's watersheds. Today many of these trees are considered serious pests. Introduced mosquitoes, rats, cats, mongoose, and exotic birds have also contributed to the extensive loss of Hawai'i's unique birdlife.

Cat eating endangered Hawaiian Gallinule.(MICHAEL WALTHER)

Mongoose. (MICHAEL WALTHER)

When we consider the history of the staggering loss of species in Hawai'i during the last millennium and how many of the original total have already succumbed to the many agents of extinction still operating in the island's forests, it is clear Hawai'i's remaining native land birds are at a high risk of being lost forever.

Maui Parrotbill. (MICHAEL WALTHER)

Of the surviving twenty-six endemic land bird species, sixteen are listed as either endangered or threatened (61.5 percent). Some of these have dangerously low populations. Thirty 'Alalā (Hawaiian Crow) were reintroduced to the Pu'u Maka'ala Natural Area Reserve from 2016 to 2020 but after severe loses, the remaining five were returned to captivity. The total captive population is now 114 and they are extinct in the wild.[1, 2] The ʻAkikiki (Kauaʻi Creeper) has declined drastically since 2016 and they are on the brink of extinction with fewer than fifty birds remaining.[3] The Kiwikiu (Maui Parrotbill) is sadly also very close to extinction with less than 150 birds surviving.[4] An attempt to establish a new population on the leeward slope of Haleakalā Crater in 2019 ended tragically when twelve of the fourteen critically endangered Kiwikiu died from avian malaria after being bitten by infected mosquitoes.[5]

ʻAkekeʻe. (MICHAEL WALTHER)

Fewer than 500 Puaiohi (Small Kauaʻi Thrush) and ʻAkekeʻe (Kauaʻi ʻĀkepa) exist and avian malaria is the most serious cause for the decline of the Akekeʻe population.[6] Unfortunately, due to global warming there is very little mosquito free habitat left on the 5,243-foot island. These extremely rare species are now found only in the wet, cold, rat and mosquito infested rainforest of the high elevation Alakaʻi Wilderness Preserve but they previously lived from the coast to the mountains island wide. On Oʻahu, the island's endemic ʻElepaio population is 1,000 and falling. Several more species have slightly higher current populations. On the Big Island of Hawaiʻi, most of the 1,900 ʻAkiapōlāʻau cling to a precarious existence in a forest on the northern flank of Mauna Kea,[7] while the remaining world population of 1,000 or fewer Palila live in a small area on the other side of the planet's tallest mountain.[8]

ʻIʻiwi. (MICHAEL WALTHER)

The beautiful, but disappearing, ʻIʻiwi is the best remaining symbol of the once spectacular Hawaiian Honeycreeper family. They no longer fly on Lānaʻi or Molokaʻi and are extremely rare on Oʻahu. Twenty-eight years ago, I photographed this iconic species on Kauaʻi with ease and considered them fairly common. Sadly, today they are very difficult to find and fast declining. The United States Fish and Wildlife Service listed the 'I'iwi as a threatened species on September 20, 2017.

The historical accounts of Hawaiian bird extinctions are full of human's lack of concern, disregard, exploitation, and selfishness for the native flora and fauna. The recurring theme is one of individuals and companies putting their own interests and profits before the needs of the other life forms that live in these isolated islands. In some cases, there was complete ignorance of the ecological consequences of their actions.

While there have been a few conservationists who expressed concern about the condition of Hawaiian forests and the plight of the native birds since the late 19th century, it wasn't until the environmental revolution in the 1960s and 1970s that government and people's attitudes and priorities began to change significantly.

The 1973 Endangered Species Act (ESA) marked a major turning point in Hawai'i's history of conservation. The landmark legislation recognized:

> "That our rich natural heritage is of esthetic, ecological, educational, recreational, and scientific value to our nation and its people." It further expressed concern that many of our nation's native plants and animals were in danger of becoming extinct.[9]

Soon after the ESA signing, federal and state biologists surveyed Hawai'i's forests, documented the present locations and populations of bird species, and wrote scientific plans to guide efforts toward their recovery. Expenses for implementing the many recommendations were very costly, and unfortunately most of the funds were never provided. But it was a beginning.

Palila. (MICHAEL WALTHER)

The state of Hawai'i created the Natural Area Reserves System in 1970, and today 38,047 preserves on five islands cover 123,810 acres.[10] In 1980, the Nature Conservancy of Hawai'i established their first Hawaiian preserve, and they now have ten preserves protecting over 200,000 acres.[11] There is an additional layer of defense provided by eight multi-agency watershed partnerships protecting 1,462,000 acres of Hawaiian forests on Kaua'i, O'ahu, Moloka'i, Lāna'i, Maui, and Hawai'i. The 38,047 acre Hakalau National Wildlife Refuge was established in

1985 to protect and manage endangered Hawaiian forest birds and their rain forest habitat.[12] Hawai'i's largest national parks, Haleakalā and Hawai'i Volcanoes, have built fences in important habitat areas to exclude feral animals.

Kaua'i 'Apapane. (MICHAEL WALTHER)

Forest bird recovery teams are working daily to save the endangered birds on Kaua'i, Maui, and Hawai'i. The American Bird Conservancy started their Hawai'i campaign in 2009, and today have programs in place to save the Nihoa Millerbird, Palila, and Kiwikiu. The San Diego Zoo, United States Fish and Wildlife Service, and the Hawai'i Department of Forestry and Wildlife operate two captive propagation centers in Hawai'i that are breeding the Hawaiian Crow, Palila, 'Akikiki, 'Akeke'e, and Kiwikiu. Invasive species task forces are working to reduce alien species threats on all the major islands.

There are many examples of grass roots efforts taking place in Hawai'i that are benefitting Hawai'i's native birds. On December 2, 2014, an Indiegogo campaign was initiated by the Kaua'i Forest Birds Recovery team with a goal to raise $10,000 to purchase twenty-five specially designed rat traps. Support was excellent and by early February 2015, they had raised $27,000! In 2021, a Go Fund Me campaign raised $11,326.00 that was donated to the University of Hawai'i Mosquito Lab to help fund Wolbachia research. The Maui Forest Birds Recovery Team conducts fund raising events, provides educational presentations and participates in annual fairs with the goal of saving Maui's endangered birds.

Kaua'i 'Elepaio. (MICHAEL WALTHER)

For the first time in the history of Hawai'i many governmental agencies, corporations, non-profit groups, scientists, and concerned citizens are working to save Hawai'i's native birds. Everyone can now look back to the regrettable extinctions that occurred since these islands were first colonized by humans to learn from past mistakes.

Unfortunately, even though the remaining native forest areas have some level of protection, most of the serious agents of extinction are still present and are causing further declines to the island's native forest birds. Wild cattle, pigs, deer, sheep and goats exist in areas of Hawaiian forests even though everyone knows the damage they have caused and are continuing to cause. Feral cats, rats and mongoose continue to consume native forest birds and their nestlings and eggs. Complete eradication of hoofed animals and predators, while very much desired by biologists and environmentalists, will most likely never happen due to the tremendous cost and lack of public and government support.

Instead, the solution being implemented is strategic fencing and exclusion of ungulates. During the last few years the fencing movement has rapidly grown. Examples of fences built to exclude feral ungulates include the 2,000 acre Kamalo/Kapualei Forest Restoration Project on the island of Hawai'i, Lāna'ihale Ungulate Fence Project on Lāna'i; a 4.5-mile-long East Alaka'i Protective Fence and the 0.3 mile Wainiha Valley fence on Kaua'i.

Red-billed Leothrix. (MICHAEL WALTHER)

On Moloka'i, a new fence in the Pu'u Ali'i Natural Area Reserve would protect approximately 484 acres of near pristine forest. This fence will be an addition to Moloka'i's other two fences built in the early 1990s: a 1.5-mile long fence that bisected the Reserve from east to west, and The Nature Conservancy's 1.25-mile long fence.[13] Many miles of fencing have been installed on O'ahu, Maui, and the Big Island of Hawai'i.

While fences are available that not only exclude feral ungulates but also rats, mongoose, cats, and even mice, they are very expensive both to build and to maintain. An excellent example of this type of fence can be seen at Ka'ena Point on O'ahu. The 2,000 foot fence was built in 2011 at a cost of $290,000 ($145.00 per foot) to protect nesting seabirds. Since the fence was completed the bird populations have increased significantly. This type of fence could be constructed to protect native forest birds but would be very expensive because of the large areas of forest that would need to be encompassed.

Japanese White-eye. (MICHAEL WALTHER)

Many important forest bird habitats are finally being protected but two of the most serious causes of extinction cannot be fenced out; the mosquitoes carrying avian malaria and pox, and the numerous introduced birds competing with the native forest birds. Researchers have been studying the diseases and their transmission for over eighty years and several promising techniques to eliminate mosquitoes have been developed recently to reduce mosquito transmitted disease in humans. Scientists have determined the incompatible insect technique is the best method to save Hawai'i's extremely rare honeycreepers. IIT will suppress populations of Culex quinquefasciatus on a landscape-level in the rainforests of Kaua'i and Maui. If deployed as planned, this will be the first time that the incompatible insect technique will be used to save endangered birds anywhere on Earth.[14] It is essential that the required funding is provided and the permit and approval processes for this extremely important work are streamlined due to the dire emergency the 'Akikiki and Kiwikiu populations are experiencing now. If there are long delays to these processes, it is possible the two critically endangered honeycreepers will become extinct in the wild.

Red-vented Bulbul. (MICHAEL WALTHER)

When you walk in the Hawaiian rain forests today, in many areas, you see mostly non-native bird species including the Japanese white-eye, Red-vented and Red-whiskered Bulbuls, Common Myna, Red-billed Leothrix, Spotted Dove, White-rumped Shama, Hwa-mei, House Finch and others. These birds compete with the native forest birds for food and carry both disease and parasites. How can we ethically remove these alien birds? How effective would any removal effort be given the extensive populations that exist? A good solution is unlikely to be found.

Only twenty-six native Hawaiian forest bird species still exist. That is 25 percent of the 103 currently known total land bird taxa that existed in the Hawaiian Islands. The number of total species will increase when new species are found or described based on subfossil remains.

White-rumped Shama. (MICHAEL WALTHER)

Many people ask why should we spend so much money to save Hawai'i's bird species when they will eventually become extinct anyway? The answers include: each species carries unique genes that we need to preserve; native birds play key roles as pollinators and insect consumers and are thus vital to the health of the forests; we have a responsibility to future generations to preserve the biodiversity that currently exists; the world becomes slightly less interesting and enjoyable every time we lose another species; birds have an inherent right to survive as beings that share our planet regardless if they serve any purpose for humanity; our species has caused all of the extensive damage and untimely extinctions, and thus we have a responsibility to end this continuing tragedy.

To save what is left of Hawai'i's native land bird species requires substantial increases in public, corporate, non-profit, and governmental awareness of their plight through education; motivation and concern for their survival; determination to save them from further extinctions; and a great increase in funding, appropriations and donations.

Anyone and everyone interested should consider: donating to the American Bird Conservancy's Hawai'i program, supporting the Nature Conservancy of Hawai'i; contributing to any crowdsourcing campaigns that help Hawai'i's native birds; asking friends and relatives to support Hawai'i's native birds; writing and calling elected representatives; and sending tax-deductible corporate donations to the NPOs listed above.

If the necessary money can be raised, all of the expensive actions in the *Recovery Plan for Hawaiian Forest Birds* could be accomplished: more fences could be constructed; all the required rat, cat, and mongoose traps could be purchased and deployed; scientists working on cures to avian malaria and pox could have the resources they need; more preserves could be established and forest areas restored; captive propagation efforts could be expanded. For the first time in Hawaiian history, we can put an end to the continuum of deplorable bird extinctions that has lasted for over 700 years.

On a sunny, blue-sky day in July 1980, I visited the wild and remote Alaka'i wilderness of Kaua'i for the first time. I had studied the colorful paintings, calls and habits of the island's endemic forest birds and was excited that I might see the very rare 'Ō'ō, 'Akialoa, 'Ō'ū, Nukupu'u, Kāma'o, or Puaiohi. We started at dawn by walking along the rim of the Kalalau Valley high above the aquamarine waters far below. The rising sun cast an orange glow on the jagged ridges that towered high above the valley floor.

Kaua'i 'Amakihi. (MICHAEL WALTHER)

After an hour we entered into the multi-shades of green, native forest, surrounded by immense tree ferns, towering koa trees, and the ubiquitous 'Ōhi'a Lehua with its beautiful scarlet flowers. After several hours of walking on the muddy trails and traversing several small streams, we had seen most of the island's common forest birds at the time including the spectacular 'I'iwi, 'Akikiki, 'Anianiau, 'Akeke'e, 'Amakihi, 'Apapane and the extremely approachable 'Elepaio. By mid-morning we reached the rain-soaked bogs and more remote forests of the plateau where we hoped to see the rare birds. For the rest of the day we hiked further, searched, listened, and checked out every bird we saw, but found none of the rare birds. We returned to our car by sunset covered in mud, very tired, but satisfied we had at least observed all of the common Kaua'i forest birds.

'Anianiau. (MICHAEL WALTHER)

Looking back to that day forty-two years ago, I now know my chances of seeing an 'Akialoa at that time were non-existent because a 1969 sighting was the last ever recorded for that species. The Nukupu'u had been even more

'Elepaio. (MICHAEL WALTHER)

impossible since the last accepted observation was in 1899! The four other rare bird species were still present in the Alaka'i as of July 1980, but they were mostly located in a very remote area that required either helicopter access or a long multi-day backpacking trip.

I returned to Kaua'i in the summer of 1994 and conducted forest bird surveys for three months in the area east of the Alaka'i Wilderness as part of an internship while attending University of California at Santa Barbara. On many enjoyable days, I hiked the remote trails recording all the birds I observed. One morning, while hiking through dense rainforest high above an untamed river, I caught a glimpse of a bird as it flew up into a nearby tree: a juvenile Puaiohi with a scalloped gray-brown chest! I grabbed my camera and managed to get a few images. Finally, I had seen and photographed one of the island's rarest birds.

Puaiohi photographed in 1994. (MICHAEL WALTHER)

In 1994, I had been surveying areas close to where the last confirmed sightings of the Kāma'o and 'Ō'ū occurred in February 1989. I had missed both of these by five and a half years.

The rare bird I wanted to see most of all was the Kaua'i 'Ō'ō. In 1983, a single male was observed in its nest tree calling for its mate from the highest branches with a ringing song that could be heard over a quarter mile away. For three days, scientists watched as the lone male flew eagerly to the top of the tree and sang loudly. After several hours of desperately trying to elicit a response he went back to gathering nest material, unaware that he was the last of his kind on Earth. No female would ever answer his futile and lonely songs. For the next four seasons he repeated his hopeless attempts by calling into the vast wilderness surrounding the nest tree. He did everything right and as best he could. In April 1987, his haunting, eerie song was heard several times. Not long afterwards, on some unknown day and without any witness, the last 'Ō'ō closed his eyes and followed his long-desired mate into the dark abyss of extinction.

A lone male Kaua'i 'Ō'ō *Moho braccatus* sings by a nest cavity to attract a mate, but all to no avail. The last known female was last seen in 1981, whereas her mate was heard calling without response for the following six years. Last observed in 1985, and heard in 1987, his passing saw the extinction of the entire family of Mohoidae, a tragic scenario unparalleled in birds in modern times.[15] (JULIAN P. HUME)

Remaining Hawaiian Native Land Bird Species

16 of the 26 (62%) remaining native land bird species are listed as endangered. Current population estimates are included below.

ENDANGERED	DECLINING
ʻIo (Hawaiian Hawk) (3,000)	Pueo (Short-Eared Owl)
ʻAlalā (Hawaiian Crow) (114)	Hawaiʻi ʻElepaio
Oʻahu ʻElepaio (1,000)	Kauaʻi ʻElepaio
Ulūlu (Nihoa Millerbird) (384)	ʻŌmao
Puaiohi (Small Kauaʻi Thrush) (500)	Hawaiʻi ʻAmakihi
ʻAinohu Kauo (Laysan Finch) (9,911)	Oʻahu ʻAmakihi
Nihoa Finch (4,475)	Kauaʻi ʻAmakihi
Palila (1,263)	ʻAnianiau
Kiwikiu (Maui Parrotbill) (150)	Maui ʻAlauahio
ʻAkiapōlāʻau (1,900)	ʻApapane
ʻAkikiki (Kauaʻi Creeper) (<50)	
ʻAlawī (Hawaiʻi Creeper) (12,500)	
ʻAkekeʻe (Kauaʻi ʻĀkepa) (<500)	
ʻĀkepa (Hawaiʻi ʻĀkepa) (14,000)	
ʻĀkohekohe (Crested Honeycreeper) (1,700)	
ʻIʻiwi (362,000; 2,500 on Kauaʻi)	

ʻElepaio. (MICHAEL WALTHER)

HISTORICAL EXTINCTIONS

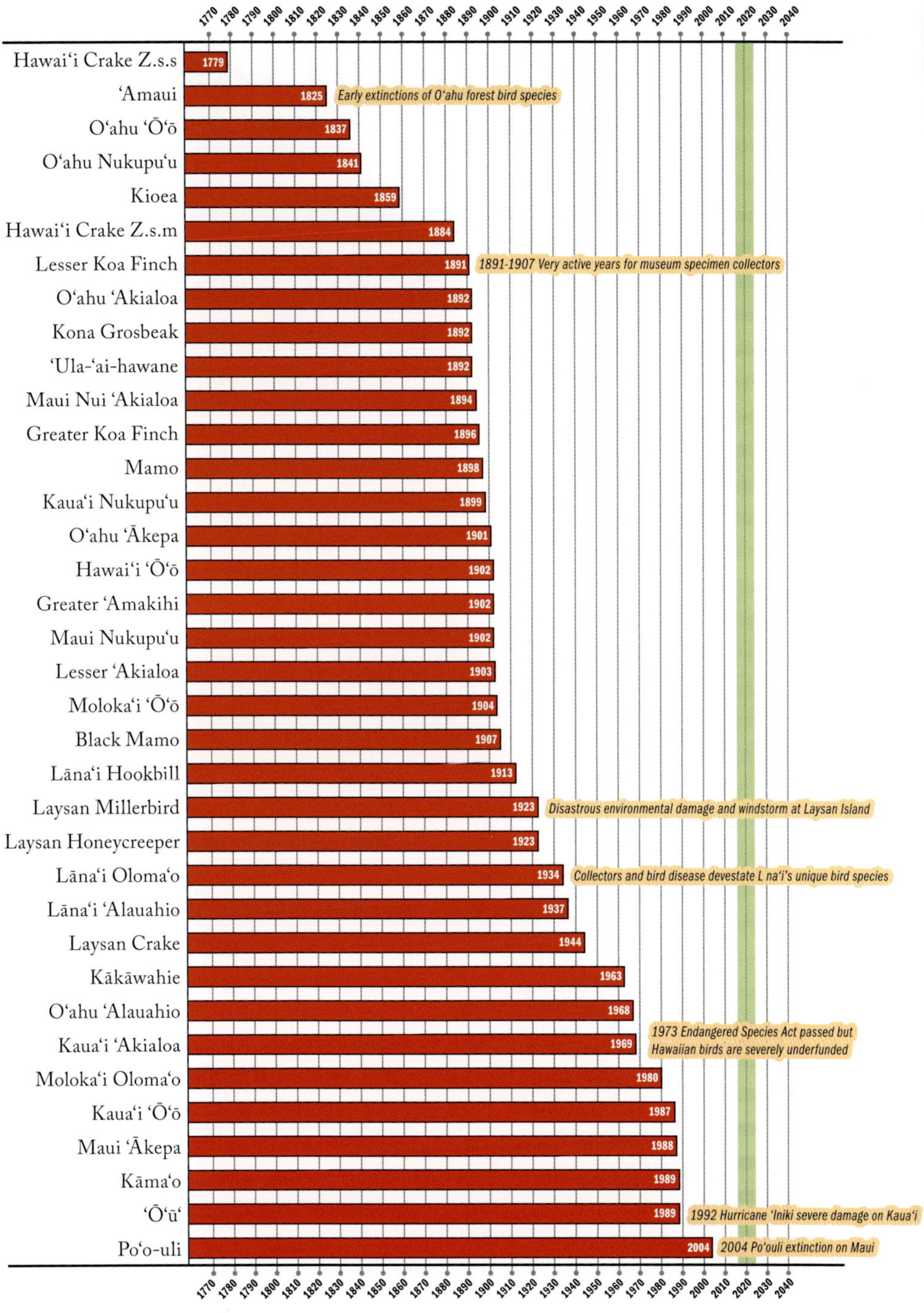

Synonymies for Extinct Hawaiian Bird Taxa

(For paintings included with each taxa)

Laysan Crake, *Zapornia palmeri*
Porzanula palmeri: Frohawk 1892 p.247

Western Hawaiian Crake, *Zapornia sandwichensis sandwichensis*
Pennula sandvicensis
Pennula sandwichensis: Sharpe, 1893a, p.252 1893b, p.443
Pennula wilsoni: Finch 1898, p.77

Eastern Hawaiian Crake, *Zapornia sandwichensis millsi*
Pennula ecaudata: Wilson & Evans, 1894, p.171
Pennula millsi: Dole, 1879, p.54

Kaua'i 'Ō'ō, *Moho braccatus*
Moho acrulocercus: Wilson & Evans, 1899, p.99

O'ahu 'Ō'ō, *Moho apicalis*
Acrulocercus apicalis: Wilson & Evans, 1894, p.103

Hawai'i 'Ō'ō, *Moho nobilis*
Acrulocercus nobilis: Wilson & Evans, 1890, p.105

Moloka'i 'Ō'ō, *Moho bishopi*
Acrulocercus bishop: Wilson & Evans, 1894, p.111

Laysan Millerbird, *Acrocephalus familiaris familiaris*
Acrocephalus familiaris: Rothschild, 1893, p.1

Lāna'i Thrush (Oloma'o), *Myadestes lanaiensis lanaiensis*
Phaeornis lanaiensis: Wilson, 1891, p.460

Kāma'o, *Myadestes myadestinus*
Phaeornis myadestina: Stejneger, 1887, p.90

Lāna'i 'Alauahio, *Paroreomyza montana montana*
Himatione montana: Wilson, 1889, p.446
Oreomyza Montana: Wilson & Evans, 1894, p.45

Kākāwahie (Moloka'i 'Alauahio), *Paroreomyza flammea*
Loxops flammea: Wilson, 1889, p.445
Oreomyza flammea: Wilson & Evans, 1899, p.39

O'ahu 'Alauahio, *Paroreomyza maculata*
Himatione maculata: Cabanis, 1850, p.100
Oreomyza maculata: Wilson & Evans, 1899, p.43

Kaua'i 'Akialoa, *Akialoa procerus*
Hemignathus procerus: Cabanis, 1890, p.331

O'ahu 'Akialoa, *Akialoa ellisiana*
Hemignathus lichtensteini: Wilson, 1889b, p.401
Hemignathus ellisianus: Gray, p.9

Maui Nui 'Akialoa, *Akialoa lanaiensis*
Hemignathus lanaiensis: Rothschild, 1893, p.24
Heterorhynchus affinis: Rothschild, 1893-1900, p.103

Lesser 'Akialoa, *Akialoa obscurus*
Hemignathus obscurus: Lichtenstein, 1839

Kaua'i Nukupu'u, *Hemignathus hanapepe*
Heterorhynchus hanapepe: Wilson & Evans, 1899, p.81

O'ahu Nukupu'u, *Hemignathus lucidus*
Heterorhynchus lucidus: Wilson & Evans, 1899, p.73

Maui Nukupu'u, *Hemignathus affinis*
Heterorhynchus affinis: Wilson & Evans, 1899, p.77

O'ahu 'Ākepa, *Loxops wolstenholmei*
Loxops rufa: Rothschild, 1895, p.54

Maui 'Ākepa, *Loxops ochraceus*
Loxops ochracea: Rotschild, 1892, p.16
Loxops aurea: Perkins, 1895, p.117

Greater Koa Finch, *Rhodacanthis palmeri*
Telespiza palmeri: Rothschild, 1900, p.201

Lesser Koa Finch, *Rhodacanthis flaviceps*
Telespiza flaviceps: Rothschild, 1900, p.201

Black Mamo, *Drepanis funerea*
Drepanorhamphus funereus: Rothschild, 1900, p.163

Endnotes

Described From Subfossils

[1] Ellen E. Paxinos, Helen F. James, Storrs L. Olson, Michael D. Sorenson, Jennifer Jackson, and Robert C. Fleischer. mtDNA from fossils reveals a radiation of Hawaiian geese recently derived from the Canada goose *(Brantacanadensis).* PNAS 2002 99: 1399-1404.

[2] Storrs L. Olson, and Helen F. James. "Prodromus of the fossil avifauna of the Hawaiian islands" *Smithsonian Contributions to Zoology*, 365, 1–59. (1982b).

[3] Ellen E. Paxinos. mtDNA from fossils reveals a radiation of Hawaiian geese recently derived from the Canada goose *(Brantacanadensis).* PNAS 2002 99: 1399-1404.

[4] Michael D. Sorenson, Alan Cooper, Ellen E. Paxinos, Thomas W. Quinn, Helen F. James, Storrs L. Olson, Robert C. Fleischer. *Relationships of the extinct moa-nalos, flightless Hawaiian waterfowl, based on ancient DNA* Proc. R. Soc. Lond. B: 1999 266 2187-2193; DOI: 10.1098/rspb. 1999.0907. Published 7 November 1999.

[5] Frank Hailer, Helen F. James, Storrs L. Olson, and Robert C. Fleischer, *Distinct and extinct: Genetic differentiation of the Hawaiian eagle*, Molecular Phylogenetics and Evolution (83 40-43, 2015).

[6] Storrs L. Olson, and Helen F. James, "Descriptions of thirty-two new species of birds from the Hawaiian Islands: Part 1. Non-Passeriformes." *Ornithological Monographs*, 45: 1-88. (1991).

[7] ibid

[8] ibid

[9] Andrew L. Iwaniuk, Storrs L. Olson, & Helen F. James, "Extraordinary cranial specialization in a new genus of extinct ducks (Aves: Anseriformes) from Kaua'i, Hawaiian Islands." *Zootaxa* 2296: 47-67.

[10] Storrs L. Olson, and Helen F. James, "Descriptions of thirty-two new species of birds from the Hawaiian Islands: Part 1. Non-Passeriformes." *Ornithological Monographs*, 45: 1-88. (1991).

[11] ibid

[12] ibid

[13] Frank Hailer, Helen F. James, Storrs L. Olson, and Robert C. Fleischer. *Distinct and extinct: Genetic differentiation of the Hawaiian eagle*, Molecular Phylogenetics and Evolution (83 40-43, 2015).

[14] Storrs L. Olson, and Helen F. James. "Descriptions of thirty-two new species of birds from the Hawaiian Islands: Part 1. Non-Passeriformes." *Ornithological Monographs*, 45: 1-88. (1991).

[15] ibid

[16] Julian P. Hume, "Undescribed juvenile plumage of the Laysan Crake or Crake *Zapornia palmeri* (Frohawk, 1892) and a detailed chronology of its extinction." *Wilson Journal of Ornithology* 129(3): 429–446. (2017)

[17] Storrs L. Olson, and Helen F. James. "Descriptions of thirty-two new species of birds from the Hawaiian Islands: Part 1. Non-Passeriformes." *Ornithological Monographs*, 45: 1-88. (1991).

[18] ibid

[19] ibid

[20] ibid

[21] ibid

[22] ibid

[23] ibid

[24] ibid

[25] Helen F. James and Storrs L. Olson, "Descriptions of thirty-two new species of birds from the Hawaiian Islands: Part 2. Passeriformes." *Ornithological Monographs*, 45: 1-86. (1991).

[26] ibid

[27] ibid

[28] ibid

[29] Olson, S. L. 2012. History, structure, evolution, behavior, distribution, and ecology of the extinct Hawaiian genus Ciridops (Fringillidae, Carduelini, Drepanidini). *Wilson Journal of Ornithology* 124(4): 651–674.

[30] Helen F. James and Storrs L. Olson, "Descriptions of thirty-two new species of birds from the Hawaiian Islands: Part 2. Passeriformes." *Ornithological Monographs*, 45: 1-86. (1991).

[31] Fleischer R.C., James H. F. & Olson S. L. 2008. Convergent evolution of Hawaiian and Australo-Pacific honeyeaters from distant songbird ancestors. *Current Biology* 18(24): 1927–1931.

[32] Julian P. Hume, Hume. *Extinct Birds*, 2nd edition. London: Helm. (2017). 608 pp.

[33] Helen F. James and Storrs L. Olson, "Descriptions of thirty-two new species of birds from the Hawaiian Islands: Part 2. Passeriformes." *Ornithological Monographs*, 45: 1-86. (1991).

[34] ibid

[35] ibid

[36] Helen F. James and Storrs L Olson, "A giant new species of Nukupuu (Fringillidae: Drepanidini: Hemignathus) from the island of Hawaii. *The Auk*, 120(4): 970-981. (2003).

[37] Helen F. James and Storrs L. Olson, "Descriptions of thirty-two new species of birds from the Hawaiian Islands: Part 2. Passeriformes." *Ornithological Monographs*, 45: 1-86. (1991).

[38] ibid

[39] ibid

[40] ibid

[41] ibid

[42] Helen F. James and Storrs L. Olson, "The diversity and biogeography of koa-finches (Drepanidini: Rhodacanthis), with descriptions of two new species." *Zoological Journal of the Linnean Society* 144: 527-541 (2005).

[43] James, H. F. & Olson, S. L. 2005. The diversity and biogeography of koa-finches (Drepanidini: Rhodacanthis) with descriptions of two new species. *Zoological Journal of the Linnean Society* 144: 527–541.

[44] Helen F. James and Storrs L. Olson, "The diversity and biogeography of koa-finches (Drepanidini: Rhodacanthis), with descriptions of two new species." *Zoological Journal of the Linnean Society* 144: 527-541. (2005).

[45] Helen F. James and Storrs L. Olson, "Descriptions of thirty-two new species of birds from the Hawaiian Islands: Part 2. Passeriformes." *Ornithological Monographs*, 45: 1-86. (1991).

[46] ibid

[47] ibid

[48] Helen F. James and Storrs L. Olson, "A new species of Hawaiian finch (Drepanidini: *Loxiodes*) from Makauwahi Cave." *Auk* ; Vol. 123 Issue 2, p335. (2006).

[49] Helen F. James and Storrs L. Olson, "Descriptions of thirty-two new species of birds from the Hawaiian Islands: Part 2. Passeriformes." *Ornithological Monographs*, 45: 1-86. (1991).

[50] ibid

[51] ibid

[52] Olson, S. L. 2012. History, structure, evolution, behavior, distribution, and ecology of the extinct Hawaiian genus *Ciridops* (Fringillidae, Carduelini, Drepanidini). *Wilson Journal of Ornithology* 124(4): 651–674.

Extinctions Since 1778

[1] Walter Rothschild, *The Avifauna of Laysan and the Neighboring Islands with a complete history to date of the birds of the Hawaiian possessions* (London: R. H. Porter, 1893-1900), 9.

[2] Rothschild, *The Avifauna*, 9.

[3] ibid

[4] Friedrich Heinrich von Kittlitz, *Nachricht von den Brüteplätzen einiger tropischen Seevögel im stillen Ocean*. Mus. Senckenbergianum 1: 115-126. 1834.

[5] Rothschild, *The Avifauna*, ix-x.

[6] Julian P. Hume, "Undescribed juvenile plumage of the Laysan Crake or Crake *Zapornia palmeri* (Frohawk, 1892) and a detailed chronology of its extinction." *Wilson Journal of Ornithology* 129(3): 429–446. (2017)

[7] ibid

[8] Rothschild, *The Avifauna*, x.

[9] Rothschild, *The Avifauna*, xi.

[10] Walter Fisher, *Birds of Laysan and the Leeward Islands, Hawaiian group.* (Washington, [D.C.]: U.S. Govt. print. off., 1903), 32.

[11] ibid

[12] Fisher, *Birds*, 33.

[13] Julian P. Hume, "Undescribed juvenile plumage of the Laysan Rail or Crake Zapornia palmeri (Frohawk, 1892) and a detailed chronology of its extinction." *Wilson Journal of Ornithology* 129(3): 429–446. (2017)

[14] Homer R. Dill and W. A. Bryan. "Report on an expedition to Laysan Island in 1911." *U.S. Dept. Agric. Biol. Surv. Bull.* 42: 1-30. (1912): 21.

[15] Dill and Bryan, Report, 9-10.

[16] Storrs Olson, "History and ornithological journals of the Tanager expedition of 1923 to the Northwestern Hawaiian Islands, Johnston and Wake Islands." *Atoll Research Bull.* 433: (1996a.): 124.

[17] Olson, History, 137.

[18] Harvey Fisher and Paul Baldwin, "War and the birds of Midway Atoll." *Condor* 48:3-1(1946): 8.

[19] John Latham, *A General Synopsis of Birds,* Vol. 1, Part 2. (London: Benjamin White, 1782), 236 https://archive.org/stream/generalsynopsi3111785lath/generalsynopsi3111785lath_djvu.txt

[20] R. L. Pyle and P. Pyle. 2017. *The Birds of the Hawaiian Islands: Occurrence, History, Distribution, and Status.* B.P. Bishop Museum, Honolulu, HI, USA. Version 2 (1 January 2017) http://hbs.bishopmuseum.org/birds/rlp-monograph

[21] Robert Kerr, *A General History and Collection of Voyages and Travels,arranged in systematic order: forming a complete history of the origin and progress of navigation, discovery, and commerce, by sea and land,from the earliest ages to the present time* (1824) Vol. XVII (Edinburg: William Blackwood and London: T. Cadell, MDCCCXXIV), 16. http://www.gutenberg.org/files/15425/15425-h/15425-h.htm

[22] Walter Rothschild, *The Avifauna of Laysan and the Neighboring Islands with a complete history to date of the birds of the Hawaiian possessions* (London: R. H. Porter, 1893-1900), 241.

[23] Scott B. Wilson and A. H. Evans, *Aves Hawaiiensis: the Birds of The Sandwich Islands* (London: R. H. Porter, 1890-99), 152.

[24] Wilson and Evans, Aves, 153.

[25] ibid

[26] Rothschild, *The Avifauna*, 242.

[27] George C. Munro, *Birds of Hawaii* 2nd. Edition. (Rutland, Vt. and Tokyo, Japan: Charles E. Tuttle Co. Inc. 1960), 144.

[28] Scott B. Wilson and A. H. Evans, *Aves Hawaiiensis: The Birds of the Sandwich Islands* (London: R. H. Porter, 1890-99), 101.

[29] ibid

[30] ibid

[31] Titian R. Peale, *United States Exploring Expedition. During the years 1838, 1839, 1840, 1841, 1842. Under the Command of Charles Wilkes, U. S. N. Vol. 8. Mammalia and Ornithology* (Philadelphia: C. Sherman, 1848), 149.

[32] Wilson and A. H. Evans, *Aves*. 102.

[33] ibid

[34] Wilson and A. H. Evans, *Aves*. 103.

[35] Walter Rothschild, *The Avifauna of Laysan and the Neighboring Islands with a complete history to date of the birds of the Hawaiian possessions* (London: R. H. Porter, 1893-1900), 228.

[36] Rothschild, *Avifauna*, 228.

[37] ibid

[38] George C. Munro, *Birds of Hawaii* 2nd. Edition. (Rutland, Vt. and Tokyo, Japan: Charles E. Tuttle Co. Inc. 1960), 84.

[39] Richardson and John Bowles, "A survey of the birds of Kauai, Hawaii." *Bull. Bernice P. Bishop Mus.* No. 227 (1964): 267.

[40] R. L. Pyle and P. Pyle. 2017. *The Birds of the Hawaiian Islands: Occurrence, History, Distribution, and Status.* B.P. Bishop Museum, Honolulu, HI, U.S.A. Version 2 (1 January 2017) http://hbs.bishopmuseum.org/birds/rlp-monograph

[41] Walter Rothschild, *The Avifauna of Laysan and the Neighboring Islands with a complete history to date of the birds of the Hawaiian possessions* (London: R. H. Porter, 1893-1900), 223.

[42] Rothschild, *Avifauna*, 223.

[43] ibid

[44] Storrs L. Olson. "The contribution of the voyage of HMS *Blonde* (1825) to Hawaiian ornithology." *Archives of Natural History*, 23 (1): 1-42. (1996): 13-14.

[45] Olson, *The Contribution*, 13-14.

[46] John K. Townsend, *Narrative of a journey across the Rocky Mountains to the Columbia River and a visit to the Sandwich Islands, Chili, &c. with a scientific appendix* (Philadelphia, PA: Henry Perkins, 1839), 269.

[47] John K. Townsend, *The Literary Record and Journal of the Linnaean Association of Pennsylvania College*, Volume 3. Gettysburg, Pennsylvania (1846): 115.

[48] Henry. W Henshaw, *Birds of the Hawaiian Islands being a complete list of the birds of the Hawaiian possessions with notes on their habits* (Honolulu: Thos. G. Thrum, 1902), 73.

[49] Scott B. Wilson and A. H. Evans, *Aves Hawaiiensis: The Birds of the Sandwich Islands* (London: R. H. Porter, 1890-99), 104.

[50] ibid

[51] R. L. Pyle and P. Pyle. 2017. *The Birds of the Hawaiian Islands: Occurrence, History, Distribution, and Status.* B.P. Bishop Museum, Honolulu, HI, U.S.A. Version 2 (1 January 2017) http://hbs.bishopmuseum.org/birds/rlp-monograph

[52] Robert Kerr, *A General History and Collection of Voyages and Travels,arranged in systematic order: forming a complete history of the origin and progress of navigation, discovery, and commerce, by sea and land,from the earliest ages to the present time* (1824) Vol. XVII (Edinburg: William Blackwood and London: T. Cadell, MDCCCXXIV), 9.

[53] Titian R. Peale, *United States Exploring Expedition. During the years 1838, 1839, 1840, 1841, 1842. Under the Command of Charles Wilkes, U. S. N. Vol. 8. Mammalia and Ornithology.* (Philadelphia: C. Sherman, 1848), 148.

[54] Scott B. Wilson and A. H. Evans, *Aves Hawaiiensis: The Birds of the Sandwich Islands* (London: R. H. Porter, 1890-99), 107-109.

[55] Wilson and A. H. Evans, *Aves*, 107-109.

[56] ibid

[57] ibid

[58] ibid

[59] Henry W. Henshaw, *Birds of the Hawaiian Islands being a complete list of the birds of the Hawaiian possessions with notes on their habits* (Honolulu: Thos. G. Thrum, 1902), 70.

[60] Henshaw, *Birds*, 71.

[61] Henshaw, *Birds*, 72.

[62] Walter Rothschild, *The Avifauna of Laysan and the Neighboring Islands with a complete history to date of the birds of the Hawaiian possessions* (London: R. H. Porter,1893-1900),Part 3. 225.

[63] Rothschild, *The Avifauna*, 225.

[64] Rothschild, *The Avifauna,* 16 (Di.).

[65] ibid

[66] Rothschild, *The Avifauna*, 225.

[67] R. C. L. Perkins, *Vertebrata*. Pp. 365-466 in D. Sharp, editor. *Fauna Hawaiiensis or the zoology of the Sandwich (Hawaiian) Isles.* Volume 1, Part IV (Cambridge, England: University Press, 1903), 442.

[68] Perkins, *Vertebrata*, 442-443.

[69] George C. Munro, *Birds of Hawaii* 2nd. Edition. (Rutland, Vt. and Tokyo, Japan: Charles E. Tuttle Co. Inc. 1960), 85-86.

[70] Laura F. Judd, *Honolulu: Sketches of Life: Social, political, and religious, in the Hawaiian Islands from 1828–1861. With a Supplementary Sketch of Events to the Present Time.* New York: Anson D. F. Randolph & Company. (1880)

[71] Francisco L. Pérez, "The Silent forest: Impact of bird hunting by prehistoric Polynesians on the decline and disappearance of native avifauna in Hawai'i." *Geographies*, 1: 192–216. (2021)

[72] Laura F. Judd, *Honolulu: Sketches of Life: Social, political, and religious, in the Hawaiian Islands from 1828–1861. With a Supplementary Sketch of Events to the Present Time.* New York: Anson D. F. Randolph & Company. (1880)

[73] Judd, L. F. 1880, "Honolulu: sketches of the life: social, political, and religious, in the Hawaiian Islands from 1828 to 1861." Reprinted 1928, *Honolulu Star-Bulletin,* Honolulu.

[74] Henshaw, H. W. 1902. "Birds of the Hawaiian Islands being a complete list of the birds of the Hawaiian possessions with notes on their habits." Honolulu: Thos. G. Thrum.

[75] Sabo, S. R. 1982. "The rediscovery of Bishop's 'O'o n Maui." *'Elepaio* 42: 69-70.

[76] Banko W. E. 1981. "History of endemic Hawaiian birds. Part I. Population histories – species accounts. Forest birds: 'Elepaio, 'O'o, and Kioea." CPSU-UH Avian History Report 7A-7B: 1-253.

[77] Scott, J. M., S. Mountainspring, F. L. Ramsey, and C. B. Kepler. 1986. "Forest bird communities of the Hawaiian Islands: their dynamics, ecology, and conservation." *Studies Avian Biol.* 9: 1-431.

[78] R. L. Pyle and P. Pyle. 2017. "The Birds of the Hawaiian Islands: Occurrence, History, Distribution, and Status." *B.P. Bishop Museum*, Honolulu, HI, U.S.A. Version 2 (1 January 2017) http://hbs.bishopmuseum.org/birds/rlp-monograph

[79] Malo, D. 1838. *Mo'olelo Hawaii (Hawaiian Antiquities).* Sheldon Dibble, Lahaina, HI. [Numerous translations including N.B. Emerson, 1951, Bishop Mus. Spec. Publ. 2: 1-278; and M.N. Chun, ed., 1996, First People's Productions, Honolulu, HI]

[80] Rooij, Willem de, Benjamin Meyer-Krahmer, and Adrienne Kaeppler. 2010. *Hawaiian Featherwork: Catalogue Raisonne of pre-1900 Feathered-God Images, Cloaks, Capes, Helmets.* Dusseldorf: Feymedia Verlagsgesellschaft.

[81] Scott B. Wilson and A. H. Evans, *Aves Hawaiiensis: the Birds of The Sandwich Islands* (London: R. H. Porter, 1890-99), 114-115.

[82] Wilson and Evans, *Aves*, 114-115.

[83] John Cassin, *United States Exploring Expedition. During the years 1838, 1839, 1840, 1841, 1842. Under the Command of Charles Wilkes, U. S. N. Mammalogy and ornithology.* (Philadelphia: J. B. Lippincott & Co., 1858), 169.

[84] Cassin, *United*, 169.

[85] Cassin, *United*, 170.

[86] Wilson and Evans, *Aves*, 115.

[87] Walter Rothschild, *The Avifauna of Laysan and the Neighboring Islands with a complete history to date of the birds of the Hawaiian possessions* (London: R. H. Porter, 1893-1900), 216.

[88] Walter Rothschild, "Descriptions of seven new species of birds from the Sandwich Islands," *Annals and Mag. Nat. Hist.*, series 6: 108-112. (1892): 109.

[89] Rothschild, Descriptions, 109.

[90] Walter Rothschild, *The Avifauna of Laysan and the Neighboring Islands with a complete history to date of the birds of the Hawaiian possessions* (London: R. H. Porter,. Part 1, 1893-1900), 2.

[91] Walter K. Fisher, "Notes on the birds peculiar to Laysan Island, Hawaiian group." Auk 20: 384-397. (1903): 391.

[92] Homer R. Dill, "Report on conditions on the Hawaiian Bird Reservation with list of the birds found on Laysan." *United States Dept. Agriculture, Biol. Survey Bull.* 42: 7-23 (1912): 22-23.

[93] William A. Bryan, "Report of an expedition to Laysan Island in 1911 : under the joint auspices of the United States Department of Agriculture and University of Iowa" *Bulletin / U.S. Department of Agriculture, Biological Survey*; no. 42 Washington, D.C.: U.S. Dept. of Agriculture, Biological Survey, (1912): 26.

[94] Bryan, Report, 27.

[95] Bryan, Report, 28.

[96] Storrs L. Olson, "History and ornithological journals of the *Tanager* expedition of 1923 to the Northwestern Hawaiian Islands, Johnston and Wake Islands." *Atoll Research Bull.* 433: (1996a.): 27.

[97] Olson, History, 127.

[98] Scott B. Wilson and A. H. Evans, *Aves Hawaiiensis: The Birds of the Sandwich Islands* (London: R. H. Porter, 1890-99), xiii.

[99] Wilson and Evans, *Aves*, xiii.

[100] Andrew J. Bloxam, *Diary of Andrew Bloxam naturalist of the "Blonde"* (Honolulu: Bernice P. Bishop Mus. Special Publ. No. 10, 1925), 43.

[101] Henry W. Henshaw, *Birds of the Hawaiian Islands being a complete list of the birds of the Hawaiian possessions with notes on their habits* (Honolulu: Thos. G. Thrum, 1902), 80.

[102] Henshaw, *Birds*, 81.

[103] R. C. L. Perkins, *Vertebrata*. Pp. 365-466 in D. Sharp, editor. *Fauna Hawaiiensis or the zoology of the Sandwich (Hawaiian) Isles.* Volume 1, Part IV (Cambridge, England: University Press, 1903), 378.

[104] Walter Rothschild, *The Avifauna of Laysan and the Neighboring Islands with a complete history to date of the birds of the Hawaiian possessions* (London: R. H. Porter, 1893-1900), 63.

[105] Rothschild, *Avifauna*, 65.

[106] ibid

[107] ibid

[108] Scott B. Wilson and A. H. Evans, *Aves Hawaiiensis: The Birds of the Sandwich Islands* (London: R. H. Porter, 1890-99), 119.

[109] Rothschild, *Avifauna*, 65.

[110] George C. Munro, *Birds of Hawaii* 2nd. Edition. (Rutland, Vt. and Tokyo, Japan: Charles E. Tuttle Co. Inc. 1960), 74.

[111] Munro, *Birds*, 74-75.

[112] William A. Bryan, *Some birds of Molokai* (Honolulu: Occas. Pap. Bernice P. Bishop Mus.4: 133-176. 1908), 81-82.

[113] Bryan, Some, 82.

[114] Bryan, Some, 83.

[115] ibid

[116] Bryan, Some, 84.

[117] ibid

[118] George C. Munro, *Birds of Hawaii* 2nd. Edition. (Rutland, Vt. and Tokyo, Japan: Charles E. Tuttle Co. Inc. 1960), 75.

[119] R. L. Pyle and P. Pyle. 2017. *The Birds of the Hawaiian Islands: Occurrence, History, Distribution, and Status.* B.P. Bishop Museum, Honolulu, HI, USA. Version 2 (1 January 2017) http://hbs.bishopmuseum.org/birds/rlp-monograph

[120] Walter Rothschild, *The Avifauna of Laysan and the Neighboring Islands with a complete history to date of the birds of the Hawaiian possessions* (London: R. H. Porter, 1893-1900), 63.

[121] Rothschild, *Avifauna*, 63.

[122] ibid

[123] ibid

[124] Scott B. Wilson and A. H. Evans, *Aves Hawaiiensis: The Birds of the Sandwich Islands* (London: R. H. Porter, 1890-99), 117.

[125] Rothschild, *Avifauna*, 20 (DI).

[126] Rothschild, *Avifauna*, 65.

[127] Allison Bryan and Alvin Seale. "Notes on the birds of Kauai" *Occas. Pap. Bernice P. Bishop Mus.* 1: 129-137(1900), 137.

[128] George C. Munro, *Birds of Hawaii* 2nd. Edition. (Rutland, Vt. and Tokyo, Japan: Charles E. Tuttle Co. Inc. 1960), 77.

[129] ibid

[130] ibid

[131] ibid

[132] Frank Richardson and John Bowles, "A survey of the birds of Kauai, Hawaii." *Bull. Bernice P. Bishop Mus.* No. 227(1964): 26.

[133] Robert L. Pyle, *American Birds, The Winter Season* December 1, 1988 - February 28, (1989): 370.

[134] R. L. Pyle and P. Pyle. 2017. *The Birds of the Hawaiian Islands: Occurrence, History, Distribution, and Status.* B.P. Bishop Museum, Honolulu, HI, U.S.A. Version 2 (1 January 2017) http://hbs.bishopmuseum.org/birds/rlp-monograph

[135] Scott B. Wilson and A. H. Evans, *Aves Hawaiiensis: The Birds of the Sandwich Islands* (London: R. H. Porter, 1890-99), 46.

[136] Wilson and Evans, Aves, 45.

[137] Walter Rothschild, *The Avifauna of Laysan and the Neighboring Islands with a complete history to date of the birds of the Hawaiian possessions* (London: R. H. Porter, 1893-1900), 118.

[138] George C. Munro, *Birds of Hawaii* 2nd. Edition. (Rutland, Vt. and Tokyo, Japan: Charles E. Tuttle Co. Inc. 1960), 107.

[139] ibid

[140] ibid

[141] Scott B. Wilson and A. H. Evans, *Aves Hawaiiensis: The Birds of the Sandwich Islands* (London: R. H. Porter, 1890-99), 40.

[142] Wilson and Evans, *Aves*, 40-41.

[143] Wilson and Evans, *Aves*, 41.

[144] Wilson and Evans, *Aves*, 40.

[145] ibid

[146] Walter Rothschild, *The Avifauna of Laysan and the Neighboring Islands with a complete history to date of the birds of the Hawaiian possessions* (London: R. H. Porter, 1893-1900), 123.

[147] Rothschild, *The Avifauna,* 123.

[148] William A. Bryan, *Some birds of Molokai* (Honolulu: Occas. Pap. Bernice P. Bishop Mus.4: 133-176. 1908), 165.

[149] Bryan, *Some*, 165.

[150] George C. Munro, *Birds of Hawaii* 2nd. Edition. (Rutland, Vt. and Tokyo, Japan: Charles E. Tuttle Co. Inc. 1960), 107.

[151] Munro, *Birds*, 107.

[152] R. L. Pyle and P. Pyle. 2017. *The Birds of the Hawaiian Islands: Occurrence, History, Distribution, and Status.* B.P. Bishop Museum, Honolulu, HI, U.S.A. Version 2 (1 January 2017) http://hbs.bishopmuseum.org/birds/rlp-monograph

[153] Scott B. Wilson and A. H. Evans, *Aves Hawaiiensis: The Birds of the Sandwich Islands* (London: R. H. Porter, 1890-99), 45.

[154] ibid

[155] ibid

[156] Wilson and Evans, *Aves*, 43.

[157] Walter Rothschild, *The Avifauna of Laysan and the Neighboring Islands with a complete history to date of the birds of the Hawaiian possessions* (London: R. H. Porter,1893-1900), Di.) 19.

[158] R. C. L. Perkins, *Vertebrata.* Pp. 365-466 in D. Sharp, editor. *Fauna Hawaiiensis or the zoology of the Sandwich (Hawaiian) Isles.* Volume 1, Part IV (Cambridge, England: University Press, 1903), 414.

[159] George C. Munro, *Birds of Hawaii* 2nd. Edition. (Rutland, Vt. and Tokyo, Japan: Charles E. Tuttle Co. Inc. 1960), 108.

[160] Pyle, R. L. and P. Pyle. 2017. *The Birds of the Hawaiian Islands: Occurrence, History, Distribution, and Status.* B.P. Bishop Museum, Honolulu, HI, USA. Version 2 (1 January 2017) http://hbs.bishopmuseum.org/birds/rlp-monograph

[161] R. C. L. Perkins, "On a new genus and species of bird of the family *Drepanididae* from the Hawaiian Islands." *Ann. Mag. Nat. Hist.*, series 9, 3 (13): 250-252 (1919): 251.

[162] Perkins, "On: 251.

[163] George C. Munro, *Birds of Hawaii* 2nd. Edition. (Rutland, Vt. and Tokyo, Japan: Charles E. Tuttle Co. Inc. 1960), 122.

[164] ibid

[165] ibid

[166] Munro, *Birds*, 122-123.

[167] Munro, *Birds*, 123.

[168] Perkins, "On" 252.

[169] Walter Rothschild, *The Avifauna of Laysan and the Neighboring Islands with a complete history to date of the birds of the Hawaiian possessions* (London: R. H. Porter, 1893-1900), 109.

[170] ibid

[171] ibid

[172] Rothschild, *The Avifauna*, 109-110.

[173] ibid

[174] Henry W. Henshaw, *Birds of the Hawaiian Islands being a complete list of the birds of the Hawaiian possessions with notes on their habits* (Honolulu: Thos. G. Thrum, 1902), 23.

[175] Henshaw, *Birds*, 43.

[176] ibid

[177] Scott B. Wilson and A. H. Evans, *Aves Hawaiiensis: The Birds of the Sandwich Islands* (London: R. H. Porter, 1890-99), 63.

[178] ibid

[179] ibid

[180] Wilson and Evans, *Aves*, 63.

[181] Wilson and Evans, *Aves*, 62.

[182] ibid

[183] ibid

[184] ibid

[185] Walter Rothschild, *The Avifauna of Laysan and the Neighboring Islands with a complete history to date of the birds of the Hawaiian possessions* (London: R. H. Porter, 1893-1900), 84.

[186] George C. Munro, *Birds of Hawaii* 2nd. Edition. (Rutland, Vt. and Tokyo, Japan: Charles E. Tuttle Co. Inc. 1960), 116.

[187] Frank Richardson and John Bowles, "A survey of the birds of Kauai, Hawaii." *Bull. Bernice P. Bishop Mus.* No. 227(1964): 30.

[188] Robert L. Pyle and Peter Pyle, "The Birds of the Hawaiian Islands: Occurrence, History, Distribution, and Status" B.P. Bishop Museum, Honolulu, HI, USA. Version 1 (31 December 2009) http://hbs.bishopmuseum.org/birds/rlp-monograph http://hbs.bishopmuseum.org/birds/rlp-monograph/pdfs/08-DREP/GAKI.pdf

[189] Scott B. Wilson and A. H. Evans, *Aves Hawaiiensis: The Birds of the Sandwich Islands* (London: R. H. Porter, 1890-99), 68.

[190] Wilson and Evans, *Aves*, 68.

[191] Walter Rothschild, *The Avifauna of Laysan and the Neighboring Islands with a complete history to date of the birds of the Hawaiian possessions* (London: R. H. Porter,1893-1900), 87.

[192] ibid

[193] Rothschild, 88.

[194] Walter Rothschild, *The Avifauna of Laysan and the Neighboring Islands with a complete history to date of the birds of the Hawaiian possessions* (London: R. H. Porter, 1893-1900), 89.

[195] ibid

[196] ibid

[197] ibid

[198] Walter Rothschild, *The Avifauna*, 90.

[199] ibid

[200] Walter Rothschild, *The Avifauna*, 15 (Di).

[201] ibid

[202] Scott B. Wilson and A. H. Evans, *Aves Hawaiiensis: The Birds of the Sandwich Islands* (London: R. H. Porter, 1890-99), 68-69.

[203] Wilson and Evans, *Aves*, 69.

[204] ibid

[205] Robert Kerr, *A General History and Collection of Voyages and Travels,arranged in systematic order: forming a complete history of the origin and progress of navigation,discovery, and commerce, by sea and land,from the earliest ages to the present time* (1824) Vol. XVII (Edinburg: William Blackwood and London: T. Cadell, MDCCCXXIV), 16. http://www.gutenberg.org/files/15425/15425-h/15425-h.htm

[206] Scott B. Wilson and A. H. Evans, *Aves Hawaiiensis: The Birds of the Sandwich Islands* (London: R. H. Porter, 1890-99), 68.

[207] Walter Rothschild, *The Avifauna of Laysan and the Neighboring Islands with a complete history to date of the birds of the Hawaiian possessions* (London: R. H. Porter, 1893-1900), 93.

[208] Rothschild, *Avifauna*, 93.

[209] R. C. L. Perkins, *Vertebrata*. Pp. 365-466 in D. Sharp, editor. *Fauna Hawaiiensis or the zoology of the Sandwich (Hawaiian) Isles.* Volume 1, Part IV (Cambridge, England: University Press, 1903), 422.

[210] Perkins, *Vertebrata*, 422-423.

[211] Perkins, *Vertebrata*, 423.

[212] Scott B. Wilson and A. H. Evans, *Aves Hawaiiensis: The Birds of the Sandwich Islands* (London: R. H. Porter, 1890-99), 82.

[213] Wilson and Evans, *Aves*, 81.

[214] Walter Rothschild, *The Avifauna of Laysan and the Neighboring Islands with a complete history to date of the birds of the Hawaiian possessions* (London: R. H. Porter, 1893-1900), 102.

[215] George C. Munro, *Birds of Hawaii* 2nd. Edition. (Rutland, Vt. and Tokyo, Japan: Charles E. Tuttle Co. Inc. 1960), 117.

[216] Pyle, R. L. and P. Pyle. 2017. *The Birds of the Hawaiian Islands: Occurrence, History, Distribution, and Status.* B.P. Bishop Museum, Honolulu, HI, U.S.A. Version 2 (1 January 2017) http://hbs.bishopmuseum.org/birds/rlp-monograph

[217] Walter Rothschild, *The Avifauna of Laysan and the Neighboring Islands with a complete history to date of the birds of the Hawaiian possessions* (London: R. H. Porter, 1893-1900), 105.

[218] \Rothschild, *The Avifauna*, 105.

[219] ibid

[220] Scott B. Wilson and A. H. Evans, *Aves Hawaiiensis: The Birds of the Sandwich Islands* (London: R. H. Porter, 1890-99), 73.

[221] Rothschild, *The Avifauna*, 106.

[222] R. C. L. Perkins, *Vertebrata*. Pp. 365-466 in D. Sharp, editor. *Fauna Hawaiiensis or the zoology of the Sandwich (Hawaiian) Isles.* Volume 1, Part IV (Cambridge, England: University Press, 1903), 427.

[223] Walter Rothschild, *The Avifauna of Laysan and the Neighboring Islands with a complete history to date of the birds of the Hawaiian possessions* (London: R. H. Porter, 1893-1900), 103.

[224] ibid

[225] ibid

[226] Rothschild, *Avifauna* Part 2, 104.

[227] ibid

[228] Pyle, R. L. and P. Pyle. 2017. *The Birds of the Hawaiian Islands: Occurrence, History, Distribution, and Status.* B.P. Bishop Museum, Honolulu, HI, U.S.A. Version 2 (1 January 2017) http://hbs.bishopmuseum.org/birds/rlp-monograph

[229] Walter Rothschild, *The Avifauna of Laysan and the Neighboring Islands with a complete history to date of the birds of the Hawaiian possessions* (London: R. H. Porter, 1893-1900), 177.

[230] ibid

[231] ibid

[232] Rothschild, *The Avifauna*, 178.

[233] Walter Rothschild, *The Avifauna of Laysan and the Neighboring Islands with a complete history to date of the birds of the Hawaiian possessions* (London: R. H. Porter, 1893-1900), 173.

[234] ibid

[235] ibid

[236] Rothschild, *Avifauna*, 174.

[237] Rothschild, *Avifauna*, (Di.) 9.

[238] Rothschild, *Avifauna*, 174.

[239] Rothschild, *Avifauna*, 175.

[240] Henry. W Henshaw, *Birds of the Hawaiian Islands being a complete list of the birds of the Hawaiian possessions with notes on their habits* (Honolulu: Thos. G. Thrum, 1902), 62.

[241] George C. Munro, *Birds of Hawaii* 2nd. Edition. (Rutland, Vt. and Tokyo, Japan: Charles E. Tuttle Co. Inc. 1960), 111.

[242] Pyle, R. L. and P. Pyle. 2017. *The Birds of the Hawaiian Islands: Occurrence, History, Distribution, and Status.* B.P. Bishop Museum, Honolulu, HI, U.S.A. Version 2 (1 January 2017) http://hbs.bishopmuseum.org/birds/rlp-monograph

[243] Walter Rothschild, *The Avifauna of Laysan and the Neighboring Islands with a complete history to date of the birds of the Hawaiian possessions* (London: R. H. Porter, 1893-1900), 203.

[244] ibid

[245] ibid

[246] Rothschild, *Avifauna*, 203.

[247] Rothschild, *Avifauna*, 203, 4 (Di.).

[248] ibid

[249] VII Rothschild, *Avifauna*, 203, 4 (Di.).

[250] George C. Munro, *Birds of Hawaii* 2nd. Edition. (Rutland, Vt. and Tokyo, Japan: Charles E. Tuttle Co. Inc. 1960), 126-127.

[251] Munro, *Birds*, 127.

[252] ibid

[253] Walter Rothschild, *The Avifauna of Laysan and the Neighboring Islands with a complete history to date of the birds of the Hawaiian possessions* (London: R. H. Porter, 1893-1900), 205.

[254] ibid

[255] ibid

[256] Rothschild, *Avifauna*, 4 (Di.).

[257] Rothschild, *Avifauna*, 205.

[258] George C. Munro, *Birds of Hawaii* 2nd. Edition. (Rutland, Vt. and Tokyo, Japan: Charles E. Tuttle Co. Inc. 1960), 127.

[259] Munro, *Birds*, 128.

[260] Scott B. Wilson and A. H. Evans, *Aves Hawaiiensis: The Birds of the Sandwich Islands* (London: R. H. Porter, 1890-99), 97.

[261] ibid

[262] ibid

[263] George C. Munro, *Birds of Hawaii* 2nd. Edition. (Rutland, Vt. and Tokyo, Japan: Charles E. Tuttle Co. Inc. 1960), 130-131.

[264] Munro, *Birds*, 131.

[265] ibid

[266] ibid

[267] Scott B. Wilson and A. H. Evans, *Aves Hawaiiensis: The Birds of the Sandwich Islands* (London: R. H. Porter, 1890-99), 87.

[268] ibid

[269] ibid

[270] James Cook, The voyages of Captain James Cook ...:With an appendix, giving an account of the present condition of the South Sea islands, &c. (London: William Smith, MDCCCXLVI (1846), 414.

[271] Wilson and Evans, *Aves*. 86.

[272] Wilson and Evans, *Aves*. 87.

[273] ibid

[274] William A. Bryan, *Some birds of Molokai* (Honolulu: Occas. Pap. Bernice P. Bishop Mus.4: 133-176. 1908), 79.

[275] R. L. Pyle and P. Pyle. 2017. *The Birds of the Hawaiian Islands: Occurrence, History, Distribution, and Status.* B.P. Bishop Museum, Honolulu, HI, USA. Version 2 (1 January 2017) http://hbs.bishopmuseum.org/birds/rlpmonograph

[276] ibid

[277] ibid

[278] ibid

[279] Scott B. Wilson and A. H. Evans, *Aves Hawaiiensis: The Birds of the Sandwich Islands* (London: R. H. Porter, 1890-99), 4.

[280] ibid

[281] James King, *A Voyage to the Pacific Ocean, Undertaken for Making Discoveries in the Northern Hemisphere: Performed Under the Direction of Captains Cook, Clerke and Gore in 1776-80*, Volume 3 (Google eBook), 119. https://play.google.com/books/reader?id=1YJcAAAAcAAJ&printsec=frontcover&output=reader&hl=en&pg=GBS.119

[282] David Malo, *Hawaiian antiquities.* 2nd. Edition (Honolulu: Bernice P. Bishop Mus. Special Pub. 2, 1951), 106.

[283] Scott B. Wilson and A. H. Evans, *Aves Hawaiiensis: The Birds of the Sandwich Islands* (London: R. H. Porter, 1890-99), 4.

[284] Walter Rothschild, *The Avifauna of Laysan and the Neighboring Islands with a complete history to date of the birds of the Hawaiian possessions* (London: R. H. Porter, 1893-1900), 160.

[285] Rothschild, *The Avifauna,* 161.

[286] ibid

[287] Rothschild, *The Avifauna*, 161-162.

[288] Henry W. Henshaw, *Birds of the Hawaiian Islands being a complete list of the birds of the Hawaiian possessions with notes on their habits* (Honolulu: Thos. G. Thrum, 1902), 51-52.

[289] ibid

[290] R. C. L. Perkins, *Vertebrata*. Pp. 365-466 in D. Sharp, editor. *Fauna Hawaiiensis or the zoology of the Sandwich (Hawaiian) Isles.* Volume 1, Part IV (Cambridge, England: University Press, 1903), 398-399.

[291] Perkins, *Vertebrata*, 399.

[292] ibid

[293] Walter Rothschild, *The Avifauna of Laysan and the Neighboring Islands with a complete history to date of the birds of the Hawaiian possessions* (London: R. H. Porter, 1893-1900), 165.

[294] ibid

[295] ibid

[296] R. C. L. Perkins, *Vertebrata*. Pp. 365-466 in D. Sharp, editor. *Fauna Hawaiiensis or the zoology of the Sandwich (Hawaiian) Isles.* Volume 1, Part IV (Cambridge, England: University Press, 1903), 402.

[297] ibid

[298] ibid

[299] ibid

[300] Alfred Newton, *On a new species of Drepanis discovered by Mr. R. C. L. Perkins* (London: Proc. of the Zoological Soc., 1893), 690.

[301] William A. Bryan, *Some birds of Molokai* (Honolulu: Occas. Pap. Bernice P. Bishop Mus. 4: 133-176. 1908), 148.

[302] ibid

[303] Bryan, *Some*, 151.

[304] Bryan, *Some*, 151-152.

[305] George C. Munro, *Birds of Hawaii* 2nd. Edition. (Rutland, Vt. and Tokyo, Japan: Charles E. Tuttle Co. Inc. 1960), 93.

[306] Walter Rothschild, *The Avifauna of Laysan and the Neighboring Islands with a complete history to date of the birds of the Hawaiian possessions* (London: R. H. Porter, 1893-1900), 3.

[307] ibid

[308] ibid

[309] ibid

[310] Rothschild, *The Avifauna*, 4.

[311] Rothschild, *The Avifauna*, X.

[312] Walter K. Fisher, "Notes on the birds peculiar to Laysan Island, Hawaiian group." *Auk* 20 (1903): 389.

[313] Fisher, "Notes" 389.

[314] Homer R. Dill, "Report on conditions on the Hawaiian Bird Reservation with list of the birds found on Laysan." United States Dept. Agriculture, Biol. Survey Bull. 42: (1912): 22.

[315] Storrs Olson, "History and ornithological journals of the *Tanager* expedition of 1923 to the Northwestern Hawaiian Islands, Johnston and Wake Islands." *Atoll Research Bull.* 433: (1996a.), 124.

[316] Olson, History, 132.

[317] Olson, History, 137.

[318] Olson, History, 139.

[319] Olson, History, 141.

[320] Olson, History, 142.

[321] ibid

[322] Olson, History, 204.

[323] Christian Melgar http://www.birdinghawaii.co.uk/XPoouli2.htm

[324] Thane K., Pratt, Cameron B. Kepler and Tonnie L. Casey. 1997. Poouli (Melamprosops phaeosoma), The Birds of North America Online (A. Poole, Ed.). Ithaca: Cornell Lab of Ornithology; Retrieved from the Birds of North America Online: http://bna.birds.cornell.edu/bna/species/272 doi:10.2173/bna.272 http://bna.birds.cornell.edu/bna/species/272/articles/measurements

[325] ibid

[326] ibid

[327] U. S. Fish and Wildlife Service. Revised Recovery Plan for Hawaiian Forest Birds. Region 1, Portland, OR. (2006): 2-147.

[328] U.S. Fish and Wildlife Service, Revised (2006).: 2-147-148.

[329] U.S. Fish and Wildlife Service, Revised, (2006): 2-148.

[330] http://www.fws.gov/news/ShowNews.cfm?ID=8F67D738-00FE-66E8-2E7A83B88192BD48

[331] Walter Rothschild, *The Avifauna of Laysan and the Neighboring Islands with a complete history to date of the birds of the Hawaiian possessions* (London: R. H. Porter, 1893-1900), 183.

[332] ibid

[333] Sanford Dole, "List of birds of the Hawaiian Islands. Corrected for the Hawaiian Annual, with valuable additions." Pages 41-58 *in Hawaiian Almanac and Annual for 1879.* (Honolulu, HI: Thomas G. Thrum., 1879): 49-50.

[334] Scott B. Wilson and A. H. Evans, *Aves Hawaiiensis: The Birds of the Sandwich Islands* (London: R. H. Porter, 1890-99), 24.

[335] Henry W. Henshaw, *Birds of the Hawaiian Islands being a complete list of the birds of the Hawaiian possessions with notes on their habits* (Honolulu: Thos. G. Thrum, 1902), 57-58.

[336] Henshaw, *Birds*, 58.

[337] Walter Rothschild, *The Avifauna of Laysan and the Neighboring Islands with a complete history to date of the birds of the Hawaiian possessions* (London: R. H. Porter, 1893-1900), 183-184.

[338] ibid

Conclusion

[1] https://www.civilbeat.org/2021/04/how-scientists-are-pivoting-in-their-quest-to-save-hawaiis-crows/ By Brittany Lyte, April 16, 2021

[2] https://science.sandiegozoo.org/species/alala (114)

[3] https://www.thegardenisland.com/2021/12/10/hawaii-news/rescued-kauai-akikiki-flown-to-maui-bird-conservation-center/

[4] https://en.wikipedia.org/wiki/Maui_parrotbill

[5] https://mauiforestbirds.org/wp-content/uploads/2020/11/20201105_2019KiwikiuConservationTranslocationReport.pdf

[6] https://www.science.org/doi/10.1126/sciadv.1600029#T1

[7] https://dlnr.hawaii.gov/wildlife/files/2019/03/SWAP-2015-Akiapolaau-Final.pdf

[8] https://science.sandiegozoo.org/news/six-palila-released-newly-restored-forest-hawaii-island

[9] http://www.fws.gov/endangered/laws-policies/

[10] http://dlnr.hawaii.gov/ecosystems/nars/

[11] http://www.nature.org/ourinitiatives/regions/northamerica/unitedstates/hawaii/

[12] http://www.fws.gov/refuges/profiles/index.cfm?id=12536

[13] http://dlnr.hawaii.gov/ecosystems/files/2013/08/Final-EA-Puu-Alii-Conservation-Fencing-Project.pdf

[14] https://www.mauinews.com/news/local-news/2021/12/biologists-try-to-beat-the-clock-to-save-mauis-endangered-birds/

[15] Julian P. Hume, Hume, *Extinct Birds*, 2nd edition. London: Helm. (2017). 608 pp.

Glossary

Adaptive radiation: a process in which organisms diversify rapidly into a multitude of new forms, particularly when a change in the environment makes new resources available, creates new challenges and opens environmental niches.

Archipelago: sometimes called an island group or island chain, it is a chain, cluster or collection of islands.

Alien: animals, plants or other organisms introduced by man into places out of their natural range of distribution, where they become established and disperse, generating a negative impact on the local ecosystem and species.

Co-evolved: the change of a biological object triggered by the change of a related object.

Culmen: the upper ridge of a bird's beak.

Deforestation: the conversion of forested areas to non-forest land use. It can be the result of the deliberate removal of forest cover for agriculture or urban development, or it can be a consequence of grazing animals, wild or domesticated.

Divergence: the evolutionary tendency or process by which animals or plants that are descended from a common ancestor evolve into different forms when living under different conditions.

Ecological: relating to or concerned with the relation of living organisms to one another and to their physical surroundings.

Ecosystem: the community of living organisms in a particular place, together with the non living physical environment in which they live.

Endemic: confined to a particular geographic area and with a specific distribution; found nowhere else.

Evolution: the process by which different kinds of living organisms are thought to have developed and diversified from earlier forms during the history of Earth. Consists of changes in the heritable traits of a population of organisms as successive generations replace one another. It is populations of organisms that evolve not individual organisms.

Fauna: all of the animal life of any particular region or time.

Flora: the plant life occurring in a particular region or time, generally the naturally occurring or indigenous native plant life.

Genus: a principal taxonomic category that ranks above species and below family, and is denoted by a capitalized Latin name.

Herbivore: an animal anatomically and physiologically adapted to eating plant material, for example foliage, for the main component of its diet. As a result of their plant diet, herbivorous animals typically have mouthparts adapted to rasping or grinding.

Infestation: to overrun in large numbers, usually so as to be harmful.

Irides: the colored part of the eye surrounding the pupil and controlling the latter's size by expanding or contracting according to light intensity.

Invasive species: introduced species (also called "non-indigenous" or "non-native") that adversely affect the habitats and bioregions they invade economically, environmentally, and/or ecologically.

Microrganism: any organism too small to be viewed by the unaided eye, as bacteria, protozoa, and some fungi and algae.

Native: an organism that originated in an area which it lives.

Naturalized: thoroughly established and replacing itself by vegetative or sexual means, but originally coming from another area. As used here, introduced, intensionally or unintentionally, by man or his activities.

Organism: an organism is any contiguous living system, such as an animal, insect, plant or bacterium. All known types of organism are capable of some degree of response to stimuli, reproduction, growth and development and self-regulation.

Pest: any animal, plant or disease that is injurious to agriculture, commerce, human health or the environment.

Phylogenetic analysis: the means of inferring or estimating relationships among individuals or groups of organisms. The evolutionary history inferred from phylogenetic analysis is usually depicted as branching, treelike diagrams that represent an estimated pedigree of the inherited relationships among molecules ("gene trees"), organisms, or both.

Sub-fossil: a term applied to the remains of a once-living organism in cases where the remains are not considered to be fully fossilized for one of two possible reasons: not enough time has elapsed since the animal died, or the conditions in which the remains were deposited were not optimal for fossilization.

Tarsus: part of the leg of a bird below the ankle.

Taxonomy: the science dealing with the description, identification, naming, and classification of organisms.

Ungulate: a hoofed mammal.

Bibliography

AB = American Birds

AOU= American Ornithological Union

Atkinson, I. A. E. "A reassessment of factors, particularly Rattus rattus L., that influenced the decline of endemic forest birds in the Hawaiian Islands." *Pacific Science* 31 (1977): 109-133.

Bailey, A. M. "Birds of Midway and Laysan islands." *Denver Museum Pictorial* No. 12. Denver, CO: Denver Museum of Natural History, 1956.

Baker, P. E. and H. Baker. "Kakawahie and O'ahu 'Alauahio." *Birds N. Am.* 503 (2000): 1-24.

Baker, P. E. "Status and distribution of the Po'ouli in the Hanawi Natural Area Reserve between December 1995 and June 1997." *Studies Avian Biology.* 22 (2001): 144-150.

Baldwin, P. H. "Birds of Hawaii National Park." *Audubon Magazine* 46 (1944): 147-154.

Baldwin, P. H. "Annual cycle, environment and evolution of the Hawaiian honeycreepers (Aves: Drepaniidae)." *Univ. Calif. Publ. Zool.* 52 (1953): 285-293.

Banko, W. E. "Rediscovery of Maui Nukupuu, Hemignathus lucidus affinis, and sighting of Maui Parrotbill, Pseudonestor xanthophrys, Kipahulu Valley, Maui, Hawaii." *Condor* 70 (1968): 265-266. [Also pp 129-132 in R. E. Warner, ed., Scientific Rep.]

Banko W. E. *History of endemic Hawaiian birds: specimens in museum collections.* Honolulu (HI): Cooperative National Park Resources Studies Unit, University of Hawaii at Manoa, Department of Botany. CPSU/UH *Avian History Report* (1979) 2.

Banko W. E. "History of endemic Hawaiian birds. Part I. Population histories – species accounts. Forest birds: 'Elepaio, 'O'o, and Kioea." CPSU-UH Avian History Report 7A-7B (1981): 1-253.

Banko W. E. "History of endemic Hawaiian birds. Part I. Population histories –species accounts. Forest birds: 'Amakihi, Creeper, Akepa, and Po'ouli." CPSU-UH Avian History Report 8B-8C (1984): 1-256.

Banko W. E. "History of endemic Hawaiian birds. Part I. Population histories – species accounts. Forest birds: 'Akialoa, Nukupu'u, and 'Akiapola'au." CPSU-UH Avian History Report 9 (1984a): 1-143.

Banko W. E. "History of endemic Hawaiian birds. Part I. Population histories – species accounts. Forest birds: Maui Parrotbill, 'O'u, Palila, Greater Koa Finch, Lesser Koa Finch, and Grosbeak Finch." CPSU-UH Avian History Report 10 (1986): 1-140.

Beaglehole, J. C., ed. *The voyage of the resolution and discovery 1776-1780.* Vol. 3 (1967), pt. 1. Cambridge University Press. Cambridge, UK.

Benson B. "Harbor sites yield rare fossils." *Honolulu Star Bull. & Advertiser.* [Beginning p.A3.] Sunday, 24 July 1977.

Berger, A. *Hawaiian Birdlife.* Honolulu: Univ. of Hawaii Press, 1972. 270 pp.

Berger, A. *Hawaiian Birdlife.* Second Ed. Honolulu: Univ. of Hawaii Press, 1981. 260 pp.

BirdLife International (BLI). IUCN Red List for birds (2016). Downloaded from http://www.birdlife.org.

BLI—*see* BirdLife International

Bloxam, A. *Of the natural history of the Sandwich Islands*. On birds, pp. 248-253 in M. Callcott, G.A.B. Byron, and R.R. Bloxam (eds.), Voyage of the H.M.S. Blonde to the Sandwich Islands in the years 1824-1825: Captain the Right Hon. Lord Byron, Commander. J. Murray, London. 1827.

Bloxam, A. J. *Diary of Andrew Bloxam naturalist of the "Blonde."* Bernice P. Bishop Mus. Special Publ. No. 10 (1925).

Brigham, W. Tufts. *Additional notes on Hawaiian feather work*. Honolulu: Bishop Museum Press (1903).

Brigham, W. Tufts. *Additional notes on Hawaiian feather work: second supplement*. Honolulu: Bishop Museum Press (1918).

Bryan, W. A., and A. Seale. *Notes on the birds of Kauai*. Occas. Pap. Bernice P. Bishop Museum (1900): 129-137.

Bryan, W. A. *Key to the birds of the Hawaiian group*. Memoirs of B.P. Bishop Mus. 1 (1901a): 259-332

Bryan, W. A. *Some birds of Molokai*. Occas. Pap. Bernice P. Bishop Mus. 4 (1908): 133-176.

Bryan, W. A. Report of an expedition to Laysan Island in 1911: under the joint auspices of the United States Department of Agriculture and University of Iowa. *Bulletin/U.S. Department of Agriculture, Biological Survey;* no. 42 Washington, D.C.: U. S. Dept. of Agriculture, Biological Survey (1912).

Burney, David A., Helen F. James, Linda Pigott Burney, Storrs L. Olson, William Kikuchi, Warren L. Wagner, Mara Burney, Deirdre McCloskey, Delores Kikuchi, Frederick V. Grady, Reginald Gage, and Robert Nishek. "Fossil Evidence for a Diverse *Biota from Kaua'i and Its Transformation since Human Arrival.*" Ecological Monographs 71, no. 4 (2001): 615-41.

Cabanis, J. *Museum Heineanum*, Vol. 1 (1850). Halberstadt: R. Frantz. 233 pp.

Casey, T. "Preliminary report on the bird life in Waihei Valley, Maui." *'Elepaio* 34 (1973): 46-50.

Cassin, J. *United States Exploring Expedition. During the years 1838, 1839, 1840, 1841, 1842. Under the Command of Charles Wilkes, U. S. N. Mammalogy and ornithology*. Philadelphia: J. B. Lippincott & Co, 1858.

Conant, S. and M. Stemmermann. "Birds in the Kipahulu District of Haleakala National Park." Pp. 67-76 In C.W. Smith, ed., Proceedings Third Conference in Natural Sciences, Hawaii Volcanoes National Park. Cooperative National Park Resources Studies Unit. University of Hawaii. Honolulu, HI, 1980.

Conant, S. "Birds of the Kalapana Extension." *Cooperative National Park Resources Studies Unit Technical Report* 36 (1980b): 1-43. University of Hawaii. Honolulu, HI.

Conant, S. "Recent observations of endangered birds in Hawaii's national parks." *'Elepaio* 41 (1981): 55-61.

Conant, S., H. D. Pratt and R. J. Shallenberger. "Reflections on a 1975 expedition to the lost world of the Alaka'i Swamp and other notes on the natural history, systematics, and conservation of Kaua'i birds." *Wilson Bulletin* 110 (1998): 1-22

Cook, J. and J. King. *A voyage to the Pacific Ocean...performed in His Majesty's Ships Resolution and Discovery in the years 1776, 1777, 1778, 1779, 1780.* 3 vols., London, 1784.

Cook, J. and J. King. *A voyage to the Pacific Ocean performed in His Majesty's Ships Resolution and Discovery in the years 1776, 1777, 1778, 1779, 1780.* Vol. 2 (1784). G. London: Nicol and T Cadell.

Dill, H. R. "Report on conditions on the Hawaiian Bird Reservation with list of the birds found on Laysan." *United States Dept. Agriculture, Biological Survey Bulletin* 42 (1912): 7-23.

Dole, S. B. "A synopsis of the birds hitherto described from the Hawaiian Islands. With notes by Sanford B. Dole, Esq., of Honolulu, Corresponding Member.' Proc. Boston Soc. Nat. Hist. 12 (1869): 294-309.

Dole, Sanford. "List of birds of the Hawaiian Islands. Corrected for the Hawaiian Annual, with valuable additions." Pages 41-58 in *Hawaiian Almanac and Annual for 1879.* Honolulu, HI: Thomas G. Thrum, 1879.

Dove, Carla J., and Storrs L. Olson. "Fossil Feathers from the Hawaiian Flightless Ibis (Apteribis Sp.): Plumage Coloration and Systematics of a Prehistorically Extinct Bird." Journal of Paleontology 85, no. 5 (September 2011): 892–897. doi:10.1666/10-133.1.

Dunmire, W. W. *Birds of the National Parks in Hawaii.* Honolulu: Hawaii Natural History Association, 1961. 36 pp.

E = 'Elepaio

Ellis, W. *An authentic narrative of a voyage performed by Captain Cook and Captain Clerke, in His Majesty's Ships Resolution and Discovery during the years 1776, 1777, 1778, 1779, 1780.* London: Robinson, Sewell and Debrett, 1782.

Elphick, C. S., D. L. Roberts, and J. M. Reed. "Estimated dates of recent extinctions for North American and Hawaiian birds." *Biol. Cons.* (in press) doi:10:1016/j.biocon.2009.11.026 [143 (2009):617-624, 2010]

Engilis, A. E., Jr. "Field notes on native forest birds in the Hanawi Natural Area Reserve, Maui." *'Elepaio* 50 (1990): 67-72

Evenhuis, N. J. *Barefoot on Lava: The journals and correspondence of naturalist R.C.L. Perkins in Hawai'i, 1892-1901.* Honolulu, HI: Bishop Museum Press, 2007. 412 pp.

Finsch, O. Ornithological letters from the Pacific. No. 1. *Ibis* 4th series, 4 (1880): 75-81.

Finch, O. "On the so-called "Sandwich Rail" in the Leiden Museum." *Notes of the Leiden Museum* 20 (1898): 77-80.

Fisher, W. K. "Notes on the birds peculiar to Laysan Island, Hawaiian group." *Auk* 20 (1903): 384-397.

Fisher, Harvey and Paul Baldwin. "War and the birds of Midway Atoll." *Condor* 48 (1946): 3-15.

Foster, J. T., E. J. Tweed, R. J. Camp, B. L. Woodworth, C. D. Adler, and T. Telfer. "Long term population changes of native and introduced birds in the Alaka'i Swamp, Kaua'i." *Conservation Biology* 18 (2004): 716-725.

Frohawk, F. W. Description of a new species of rail from Laysan Island (North Pacific). *Ann. Mag. Nat. Hist.* (6 Ser.) 9 (1892): 247-249.

Gorresen, P. M., R. J. Camp, M. H. Reynolds, B. L. Woodworth, and T. K. Pratt. Status and trends of native Hawaiian songbirds. Pp. 108-136 in T. K. Pratt, et al., eds., Conservaion biology of Hawaiian forest birds. New Haven, CT: Yale University Press, 2009.

Grant, P. R. "In remembrance: Chloridops kona, Died ca. 1895." *The Linnean* 2 (1995): 14-23.

Gray, G. R. *Catalogue of the birds of the tropical islands of the Pacific Ocean in the collection of the British Museum.* Trustees of the British Museum, London, 1859. 72 pp.

Gregory, H. E. "Report for the director for 1923: Ornithology." *B.P. Bishop Mus. Bull.* 10 (1924): 12-13.

Gregory, H. E. "Report of the director for 1927." *B.P. Bishop Mus. Bull.* 57 (1928): 1-38.

Gregory, H. E. "Report of the director for 1928." *B.P. Bishop Mus. Bull.* 65 (1929): 1-58

Gregory, H. E. "Report of the director for 1931." *B.P. Bishop Mus. Bull.* 94 (1932): 1-54.

Gregory, H. E. "Report of the director for 1932." *B.P. Bishop Mus. Bull.* 106 (1933): 1-49.

Gregory, H. E. "Report of the director for 1934." *B.P. Bishop Mus. Bull.* 133 (1935): 1-71.

Gregory, H. E. "Report of the director for 1935." *B.P. Bishop Mus. Bull.* 140 (1936): 1-52.

Hailer, Frank, Helen F. James, Storrs L. Olson, and Robert C. Fleischer. Distinct and extinct: Genetic differentiation of the Hawaiian eagle, *Molecular Phylogenetics and Evolution,* 83 (2015): 40-43.

Hart, A. "Hawai'i's last honeyeater." *Natl. Parks and Conservation. Mag.* Dec. 1978 (1978): 12- 26.

Henshaw, H. W. *Birds of the Hawaiian Islands being a complete list of the birds of the Hawaiian possessions with notes on their habits.* Honolulu: Thos. G. Thrum, 1902.

HFW = Hawaii's Forest and Wildlife

Hirai, L. T. "Native birds of Lanai, Hawaii." *Western Birds* 9 (1978): 71-77.

Hume, Julian P. and Michael Walters. *Extinct Birds.* London: T&AD Poyser, an imprint of Bloomsbury Pub, 2012.

Hume, Julian P. *Extinct Birds.* Second edition. London: Christopher Helm; an imprint of Bloomsbury Pub, 2017.

HWN = Hawai'i Wildlife Newsletter

Iwaniuk, Andrew L., Storrs L. Olson, & Helen F. James. "Extraordinary cranial specialization in a new genus of extinct ducks (Aves: Anseriformes) from Kaua'i, Hawaiian Islands." *Zootaxa* 2296 (2009): 47-67.

James, Helen F. "A late Pleistocene avifauna from the island of Oahu, Hawaiian Islands." in *L'Évolution des oiseaux d'après le témoignage des fossiles. Table Ronde internationale de CNRS, Lyon-Villeurbanne, 18-21 Sept. 1985. Documents des Laboratoires de Géologie de Lyon, no. 99,* edited by Mourer-Chauviré, Cécile., 1987, 121–128.

James, Helen F., and Storrs L. Olson. "Descriptions of thirty-two new species of birds from the Hawaiian Islands: Part II." Passeriformes. *Ornithological Monogr.* No 46 (1991).

James, Helen F. and Storrs L. Olson. "A giant new species of Nukupuu (Fringillidae: Drepanidini: Hemignathus) from the island of Hawaii." *The Auk*, 120(4, 2003): 970-981.

James, Helen F. "The osteology and phylogeny of the Hawaiian finch radiation (Fringillidae: Drepanidini), including extinct taxa." *Zoological Journal of the Linnean Society*. 141 (2004): 207–255.

James, Helen F. and Storrs L. Olson. "The diversity and biogeography of koa-finches (Drepanidini: Rhodacanthis), with descriptions of two new species." *Zoological Journal of the Linnean Society* 144 (2005): 527-541.

James, Helen F. and Storrs L. Olson. "A new species of Hawaiian finch (Drepanidini: *Loxiodes*) from Makauwahi Cave." *Auk*; Vol. 123 (2006) Issue 2, p 335.

James, Helen and Jonathan Price. "Integration of palaeontological, historical, and geographical data on the extinction of koa-finches." Diversity and Distributions. 14 (2008) 441 - 451. 10.1111/j.1472-4642.2007.00442.x.

Judd, L. F. "Honolulu: sketches of the life: social, political, and religious, in the Hawaiian Islands from 1828 to 1861". Reprinted 1928, *Honolulu Star-Bulletin*, Honolulu, 1880.

Kepler, C. B. and J. M. Scott. "Native forest birds on Lana'i and Moloka'i." P. 179 in C. W. Smith, ed., *Proceedings First Conference in Natural Sciences, Hawaii Volcanoes National Park. Cooperative National Park Resources Studies Unit.* University of Hawaii. Honolulu, HI (1980).

Kerr, Robert. *A General History and Collection of Voyages and Travels,arranged in systematic order: forming a complete history of the origin and progress of navigation,discovery, and commerce, by sea and land,from the earliest ages to the present time* Vol. XVII Edinburg: William Blackwood and London: T. Cadell, MDCCCXXIV (1824). http://www.gutenberg.org/files/15425/15425-h/15425-h.htm

Latham, John. *A General Synopsis of Birds*, Vol. 1 (1782), Part 2. London: Benjamin White. https://archive.org/stream/generalsynopsi3111785lath/generalsynopsi3111785lath_djvu.txt

Lichtenstein, M. H. K. *Beitrag zur ornithologischen Fauna von Californien nebst Bemerkungen über die Artkennzeichender Pelicane und über einige Vögel von den Sandwich-Inseln.* Abh. Koniglichen Akad. Wissensch. Berlin: 417-451 (1839).

Malo, D. *Mo'olelo Hawaii (Hawaiian Antiquities).* Sheldon Dibble, Lahaina, HI (1838). [Numerous translations including N.B. Emerson, 1951, Bishop Mus. Spec. Publ. 2: 1- 278; and M.N. Chun, ed., 1996, First People's Productions, Honolulu, HI]

Medway, D. G. "The contribution of Cook's third voyage to the ornithology of the Hawaiian Islands." *Pacific Science* 35 (1981): 1-175.

Melgar, Christian. http://www.birding-hawaii.co.uk/XPoouli2.htm

Munro, G. C. *The Birds of Hawaii.* Honolulu, HI: Tongg Publishing Company, 1944. Reprinted 1960 Bridgeway Press, Rutland, VT.

Munro, G. C. *Birds of Hawaii.* 2nd. Edition. Rutland, Vt. and Tokyo, Japan: Charles E. Tuttle Co., Inc, 1960.

Munro, G. C. *The story of Lana'i.* Honolulu, HI: Native Books, 2007. 233 pp.

Newton, A. "On a new species of *Drepanis* discovered by Mr. R. C. L. Perkins." *Proc. of the Zoological Soc.* London, 690 (1893).

Olson, Storrs L. and Helen F. James. *Prodromus of the fossil avifauna of the Hawaiian islands. Smithsonian Contributions to Zoology*, 365, 1–59.

Olson, Storrs L. and Helen F. James. 1991. *Descriptions of thirty-two new species of birds from the Hawaiian Islands: Part 1. Non-Passeriformes. Ornithological Monographs,* 45 (1982): 1-88.

Olson, Storrs L. and Helen F. James. "A specimen of Nuku pu'u (Aves: Drepanidini: Hemignathus lucidus) from the island of Hawai'i." *Pacific Science.* 48 (4; 1994): 331–338.

Olson, Storrs L. "The contribution of the voyage of H.M.S. Blonde (1825) to Hawaiian ornithology." Archives of Natural History 23 (1996a): 1-42.

Olson, Storrs L. "History and ornithological journals of the *Tanager* expedition of 1923 to the Northwestern Hawaiian Islands, Johnston and Wake Islands." *Atoll Research Bull.* 433, 1996b.

Olson, Storrs L. "Kona Grosbeak, Greater Koa-Finch, and Lesser Koa-Finch." *Birds of N. Am.* 424 (1999b): 1-20.

Olson, Storrs L. "History, Structure, Evolution, Behvior, Distribution, and Ecology of the Extinct Hawaiian Genus Ciridops (Fringillidae, Carduelini, Drepanidini)." *Wilson Journal of Ornithology* 124 (4; 2012): 651–674.

Paxinos, Ellen E., Helen F. James, Storrs L. Olson, Michael D. Sorenson, Jennifer Jackson, and Robert C. Fleischer. "mtDNA from fossils reveals a radiation of Hawaiian geese recently derived from the Canada goose (*Brantacanadensis*)" PNAS 99 (2002): 1399-1404.

Paxton, E. H., R. J. Camp, P. M. Gorresen, L. H. Crampton, D. L. Leonard, E. A. VanderWerf. "Collapsing avian community on a Hawaiian island." *Sci. Adv.* 2 (2016), e1600029

Peale, T. R. *United States Exploring Expedition. During the years 1838, 1839, 1840, 1841, 1842. Under the Command of Charles Wilkes, U. S. N. Vol. 8. Mammalia and Ornithology.* Philadelphia: C. Sherman, 1848.

Pekelo, N. Jr. "Nature notes from Molokai." *Elepaio* 24 (1963): 17-18.

Pekelo, N. Jr. "Nature notes from Molokai." *Elepaio* 24 (1964): 46-48.

Pérez, F. L. "The Silent Forest: Impact of Bird Hunting by Prehistoric Polynesians on the Decline and Disappearance of Native Avifauna in Hawai'i." *Geographies* 2021, 1: 192–216. https://doi.org/10.3390/geographies1030012

Perkins, R. C. L. "Notes on collecting in Kona, Hawaii." *Ibis* (6th ser.; 1893) 5: 101-112.

Perkins, R. C. L. "Notes on some Hawaiian Birds." *Ibis* (7th ser.; 1895) 1: 117-129.

Perkins, R. C. L. "Vertebrata." Pp. 365-466 in D. Sharp, editor. *Fauna Hawaiiensis or the zoology of the Sandwich (Hawaiian) Isles.* Volume 1, Part IV. Cambridge, England: University Press, 1903.

Perkins, R. C. L. (Introduction). Pp. xv-ccxxvii in D. Sharp, editor. *Fauna Hawaiiensis or the zoology of the Sandwich (Hawaiian) Isles.* Volume 1, Part VI. Cambridge, England: University Press, 1913.

Perkins, R. C. L. "On a new genus and species of bird of the family Drepanididae from the Hawaiian Islands." *Annals and Mag. Nat. Hist.*, series 9, 3 (1919): 250-252.

Pratt, H. D. "Systematics of the Hawaiian "Creepers" *Oreomystis* and *Paroreomyza*." *The Condor*, 1992, Volume 94, Issue 4, 1, Pages 836–846, https://doi.org/10.2307/1369281

Pratt, H. D. "The Hawaiian honeycreepers: Drepanidinae." *Bird Families of the World*. Oxford, England: Oxford University Press, 2005.

Pratt, H. D. "A consensus taxonomy for the Hawaiian honeycreepers." *Occasional Papers of the Museum of Natural Science*, Louisiana State University, No. 85, 2014. Available online at http://sites01.lsu.edu/wp/mnspapers/files/2014/10/85.pdf.

Pratt, H. D. and I. L. Brisbin, Jr. "Hawaiian Coot *(Fulica alai)*," version 1.0. In Birds of the World (A. F. Poole and F. B. Gill, Editors). Cornell Lab of Ornithology, Ithaca, NY, USA, 2020. https://doi.org/10.2173/bow.hawcoo.01

Pratt, Thane K. "Plant communities and bird distribution on east Molokai." *Elepaio* 33 (1973): 66-70.

Pratt, Thane K., Cameron B. Kepler and Tonnie L. Casey. Poouli (Melamprosops phaeosoma), *The Birds of North America Online* (A. Poole, Ed.). Ithaca: Cornell Lab of Ornithology, 1997. Retrieved from the Birds of North America Online: http://bna.birds.cornell.edu/bna/species/272 doi:10.2173/bna.272 http://bna.birds.cornell.edu/bna/species/272/articles/measurements

Pratt, T. K., S. G. Fancy, and C. J. Ralph. "Akiapolaau (Hemignathus wilsoni), version 1.0." In *Birds of the World* (A. F. Poole, Editor). Cornell Lab of Ornithology, Ithaca, NY, USA, (2020).

Pratt, T. K., S. G. Fancy, and C. J. Ralph. "'Akiapola'au and Nukupu'u." *Birds of N. Am.* 600 (2001): 1-28.

Pratt, Thane K. and R. L. Pyle. "Nukupu'u in the twentieth century: endangered species or phantom presence?" *'Elepaio* 60 (2000): 35-41.

Pyle, R. L. and C. J. Ralph. "Hawaiian Islands region." *Amer. Birds* 36 (1982): 221-223.

Pyle, Robert L. Volume 37 Issue 5. (September-October 1983) "The spring migration: Hawaiian Islands region." *North American Birds* 37: 844-917. {Kaua'i O'o'a'a page 915} https://sora.unm.edu/node/2628

Pyle, Robert L. Volume 41 Issue 3. (March-May 1987) "The spring migration: Hawaiian Islands region." *North American Birds* 41: 401-497 {Kaua'i O'o'a'a page 492} https://sora.unm.edu/node/114473

Pyle, Robert L. *North American Birds*, Volume 43, Number 3 (1989).

R. L. Pyle, and P. Pyle. "The Birds of the Hawaiian Islands: Occurrence, History, Distribution, and Status." *B. P. Bishop Museum*, Honolulu, HI, U.S.A., 2009. Version 2 (1 January 2017) http://hbs.bishopmuseum.org/birds/rlp-monograph

R. L. Pyle, and P. Pyle. "The Birds of the Hawaiian Islands: Occurrence, History, Distribution, and Status." *B. P. Bishop Museum*, Honolulu, HI, U.S.A. Version 2 (1 January 2017) http://hbs.bishopmuseum.org/birds/rlp-monograph

Reynolds, M. H., T. J. Snetsinger, and C. M. Herrmann. "Kauai's endangered solitaires: Update on population status and distribution." 1997b, 1996. Trans. W Section Wildl. Soc. 33: 49-55.

Reynolds, M. H. and T. Snetsinger. 2001. "The Hawaii rare bird search 1994-1996." *Studies Avian Biol.* 22: 133-143.

Richards, L. P., and P. H. Baldwin. "Recent records of some Hawaiian honeycreepers." *Condor* 55 (1953): 221-222. [condensed version reprinted E 14: 28-29] _____ and W. J. Bock. 1973. "Functional anatomy and adaptive evolution of the feeding apparatus in the Hawaiian honeycreeper genus Loxops (Drepanididae)." *Ornith. Monogr.* 15: 1-173.

Richardson, Frank. "The status of native land birds on Molokai, Hawaiian Islands." *Pacific Science* 3 (1949): 226-230.

Richardson, Frank and John Bowles. "A survey of the birds of Kauai, Hawaii." *Bull. Bernice P. Bishop Mus.* No. 227, 1964.

Roberts, D. L., C. S. Elphick, and M. L. Reid. "Identifying anomalous reports of putatively extinct species and why it matters." *Conserv. Biol.* 24 (2010): 189-196.

Rooij, Willem de, Benjamin Meyer-Krahmer, and Adrienne Kaeppler. *Hawaiian Featherwork: Catalogue Raisonne of pre-1900 Feathered-God Images, Cloaks, Capes, Helmets.* Dusseldorf: Feymedia Verlagsgesellschaft, 2010.

Rothschild, W. "Types of three new Hawaiian birds." *Bull. Brit. Brit. Ornith. Club* 1 (1892): 16.

Rothschild, W. "On H*emignathus lanaiensis*." *Bull. Brit. Ornith. Club* 1 (1893): 24-25. [Reprinted] *Ibis* 6 (5th Ser.): 256-257.

Rothschild, W. "Descriptions of three new birds from the Sandwich Islands." *Ibis* 6 (5th Ser.; 1893): 112-114.

Rothschild, W. "Note on the *Loxops* of O'ahu." *Novit. Zool.* 2 (1895): 54

Rothschild, W. *The avifauna of Laysan and the neighboring islands with a complete history to date of the birds of the Hawaiian possessions.* London: R. H. Porter, 1893-1900.

Sabo, S. R. "The rediscovery of Bishop's 'O'o n Maui." *'Elepaio* 42 (1982): 69-70.

Schauinsland, H. H. "Three months on a coral island (Laysan)." Verlag von Max Nossler. Bremen. [English translation by M. D. F. Udvardy, Ed, 1996. *Atoll Res. Bull.* 432 (1899): 1-53].

Sclater, P. L. "On the birds collected in the Sandwich Islands." Pp. 93-99 in C. W. Thomson, ed., "Report on the scientific results of the voyage of the H.M.S. *Challenger* during the years 1873-1876." Vol II. Zoology. Neill and Company, Edinburgh, Scotland, 1881.

Scott, J. M. and J. L. Sincock. "Recent observations on the birds of the Koolau Forest Reserve, Maui." *Western Birds* Vol 8, 3 (1977): 113-116.

Scott, J. M., C. B. Kepler, and J. L. Sincock. "Distribution and abundance of Hawai'i's endemic land birds: Conservation and management strategies." Pp. 75-104 in C. P. Stone and J. M. Scott, eds., "Hawai'i's terrestrial ecosystems preservation and management." Cooperative National Park Resources Studies Unit, University of Hawaii, Honolulu, HI, 1985.

Scott, J. M. and C. B. Kepler. "Distribution and abundance of Hawaiian native birds: status report." *Bird Conservation* 2 (1985): 43-70.

Scott, J. M., S. Mountainspring, F. L. Ramsey, and C. B. Kepler. "Forest bird communities of the Hawaiian Islands: their dynamics, ecology, and conservation." *Studies Avian Biol.* 9 (1986): 1-431.

Scott, J. M., F. L. Ramsey, M. Lammertink, K. V. Rosenberg, R. Rohrbaugh, J. A. Wiens, and J. M. Reed. "When is an "extinct" species really extinct? Gauging the search efforts for Hawaiian forest birds and the Ivory-billed Woodpecker." *Avian Conservation and Ecology* 3 (2008). http://www.ace-eco.org/vol3/iss2/art3/

Shallenberger, R. J. "Avifaunal survey of North Halawa Valley, Oahu." Appendix G. Fauna [Pp 1-42] In Draft supplement to the Interstate Route H-3 environmental impact statement. Report FHWA HI-EIS-77-01-D(S). Federal Highway Administration, State of Hawaii, Dept. of Transportation. Honolulu, HI, 1977.

Shallenberger, R. J. and H. D. Pratt. "Recent Observations and field identification of the Oahu Creeper." *'Elepaio* 38 (1978): 135-140.

Shallenberger, R. J. and G. K. Vaughn. "Avifaunal survey in the Central Ko'olau Range, O'ahu." Kailua, HI: Ahuimanu Productions, 1978. 106 pp.

Sharpe, R. B. "On Pennula and its allies." *Bull. Brit. Ornith. Club* 1 (1893): 19-20. Reprinted *Ibis* 6 (5th Ser.): 252-253.

Sharpe, R. B. "On Pennula sandwichensis." *Bull. Brit. Ornith. Club* 1 (1893b): 42-43. Reprinted *Ibis* 6 (5th Ser.): 443-444.

Simon, J. C., T. K. Pratt, K. E. Berlin, J. R. Kowalsky, S. G. Fancy, and J. S. Hatfield. "Temporal variation in bird counts within a Hawaiian rainforest." *Condor* 104 (2002): 469-481.

Snetsinger, T. J., M. H. Reynolds, and C. M. Herrmann. "'O'u and Lana'i Hookbill." *Birds of N. Am.* 335-336 (1998): 1-20.

Sorenson, Michael D., Alan Cooper, Ellen E. Paxinos, Thomas W. Quinn, Helen F. James, Storrs L. Olson, Robert C. Fleischer. "Relationships of the extinct moa-nalos, flightless Hawaiian waterfowl, based on ancient DNA." *Proc. R. Soc. Lond. B:* 1999 266 2187-2193; DOI: 10.1098/rspb.1999.0907. Published 7 November 1999.

Stejneger, L. "Birds of Kauai Island, Hawaiian Archipelago, collected by Mr. Valdemar Knudsen, with descriptions of new species." *Proceedings of the United States National Museum.* 10 (1887): 75-102.

Stresemann, E. "Birds collected during Capt. James Cook's last expedition (1776- 1780)." *Auk* 67 (1950): 66-88.

Sykes, P. W., Jr., A. K. Kepler, C. B. Kepler, and J. M.Scott. "Kauai'i 'O'o." *Birds of N. Am.* 535 (2000): 1-32.

Townsend, J. K. "Narrative of a journey across the Rocky Mountains to the Columbia River and a visit to the Sandwich Islands, Chili, &c. with a scientific appendix." Philadelphia: Henry Perkins, PA, 1839.

Townsend, J. K. *The Literary Record and Journal of the Linnaean Association of Pennsylvania College,* Volume 3 Gettysburg, Pennsylvania, 1846.

United States Fish and Wildlife Service. *Hawaii forest bird recovery plan:* Hawai'i Creeper, Hawai' i 'Akepa, 'Akiapola'au, 'O'u (J. M. Scott et al., eds.). USFWS, Portland, OR., 1982. 76 pp.

United States Fish and Wildlife Service. *Hawaii forest bird recovery plan: Hawai'i Creeper, Hawai'i 'Akepa, 'Akiapola'au, 'O'u* (J. M. Scott et al., eds.). USFWS, Portland, OR, 1982c. 76 pp.

United States Fish and Wildlife Service. *Hawai'i forest birds recovery plan* (J. M. Scott et al., eds.). USFWS, Portland, OR, 1983b.

United States Fish and Wildlife Service. *Kaua'i forest birds recovery plan* (J. L. Sincock et al., eds.). USFWS, Portland, OR, 1983c. 69 pp.

United States Fish and Wildlife Service. *Hawai'i forest birds recovery plan* (J. M. Scott et al., eds.). USFWS, Portland, OR, 1983d.

United States Fish and Wildlife Service. *Maui-Moloka'i forest birds recovery plan* (C. B. Kepler et al., eds.). USFWS, Portland, OR, 1984. 110 pp.

United States Fish and Wildlife Service. *Maui-Moloka'i forest birds recovery plan* (C. B. Kepler et al., eds.). USFWS, Portland, OR, 1984d. 110 pp.

United States Fish and Wildlife Service. *Revised Recovery Plan for Hawaiian Forest Birds.* Region 1, Portland, OR, 2006. http://ecos.fws.gov/docs/recovery_plan/060922a.pdf

United States Fish and Wildlife Service. *Oloma'o or Moloka'i thrush (Myadestes lanaiensis rutha). 5-Year Review Summary and Evaluation.* USFWS, Honolulu, HI, 2009.

van Riper, C. III. "Censuses and breeding observations of the birds on Kohala Mountain, Hawaii." *Wilson Bull.* 94 (1982a): 463-476.

_____. "Avifauna of Kohala Mountain, Hawaii." CPSU-UH Technical report 44 (1982b): 1-27.

VanderWerf, E. A. *Hawaiian Bird Conservation Action Plan.* Pacific Rim Conservation, Honolulu, HI, 2013.

Von Kittlitz, F. H. Nachricht von den Brüteplätzen einiger tropischen Seevögel im stillen Ocean. Mus. Senkenbergianum Abhandlungen aus dem Gebiete der beschreibenden Naturgeschite 1 (1834): 115-126.

Wakelee, K. M. and S. G. Fancy. "'Oma'o, Oloma'o, Kama'o, 'Amaui." *Birds of N. Am.* 460 (1999): 1-28.

Wilson, S. B. "Descriptions of some new species of Sandwich-Island birds." *Proc. Zool. Soc. of London* 6 (1889): 445-447.

Wilson, S. B. and A. H. Evans. *Aves Hawaiiensis: the birds of the Sandwich Islands.* London: R. H. Porter, 1890-99.

Wilson, S. B. "On some of the birds of the Sandwich Islands." *Ibis* 6th Ser. 2 (1890a): 170-196.

Wilson, S. B. "Descriptions of two new species of Sandwich-Island birds." *Annals and Magazine of Natural History* 6 (7; 1891): 460. http://www.fws.gov/news/ShowNews.cfm?ID=8F67D738-00FE-66E8-2E7A83B88192BD48

Environmental Non-Profit Organizations That Help Native Birds in Hawai'i

American Bird Conservancy
www.abcbirds.org
(888) 247-3624

Bernice Pauahi Bishop Museum
www.bishopmuseum.org
(808) 847-3511

BirdLife International
The David Attenborough Building
Pembroke Street
CAMBRIDGE
CB2 3QZ
United Kingdom
www.birdlife.org/worldwide/support-us/donate
www.justgiving.com/specieschampions/

Hawai'i Audubon Society
www.hawaiiaudubon.org
(808) 528-1432

Kaua'i Forest Bird Recovery Project
PO Box 27
Hanapepe, HI 96716
http://kauaiforestbirds.org/
(808) 335-5078

Maui Forest Bird Recovery Project
2465 Olinda Road
Makawao, HI 96768-7138
(808) 573-0280
info@mauiforestbirds.org

Nature Conservancy of Hawai'i
www.nature.org/ourinitiatives/regions/northamerica/united states/hawaii
(808) 545-2019

Pacific Rim Conservation
PO Box 61827
Honolulu, HI 96839
www.pacificrimconservation.org
(808) 377-7114

San Diego Zoo Global's Hawai'i Endangered Bird Conservation Program
PO Box 120551
San Diego, CA 92112-0551
(619) 557-3914
wildlifeconservancy@sandiegozoo.org

About the Author

Michael Walther was born in Cleveland, Ohio and raised in Margate, Florida and Southern California. He first visited Hawai'i in December 1972 for several months. In March 1980, he moved to Kaua'i and while living in Po'ipū Beach became interested in the native honeycreepers that survived in the remote Alaka'i Wilderness area. In January 1981, he returned to the mainland.

For the next fifteen years he lived and worked in California but the memory of the colorful and disappearing honeycreepers stayed with him. During his last year in college, the fate of Hawai'i's native birds became the focal point of his research. In 1994, he returned to Hawai'i which had become "the Endangered Species Capital of the World," to study its birds and do what he could to help them survive. For three months, he surveyed the native forest bird populations on Kaua'i to determine their status. His results were published in the journal of the Hawai'i Audubon Society: *'Elepaio.*

He returned to California to finish his degree in Anthropology with an emphasis in Environmental Studies from the University of California at Santa Barbara. After graduating with honors in March 1995, he moved to Maui and volunteered for The National Biological Survey. He participated in a project studying several of the rarest birds on Earth, including the Maui 'Ākepa, *Loxops coccineus ochraceous*, Maui Nukupu'u, *Hemignathus lucidus affinis,* Po'o-uli, *Melamprosops phaeosoma*, Kiwikiu, *Pseudonestor xanthophrys,* and 'Ākohekohe, *Palmeria dolei*. He worked on a research team that was surveying the wild, wet and windy slopes of the world's largest dormant volcano, Haleakalā.

Michael is the author of four books on native Hawaiian flora and fauna, including: *A Pocket Guide to Nature on O'ahu; Images of Natural Hawai'i: A Pictorial Guide of the Aloha State's Native Forest Birds and Plants; Pearls of Pearl Harbor and the Islands of Hawai'i* and *A Guide to Hawai'i's Coastal Plants,* as well as many articles in birding and nature publications.

In the fall of 1995, Michael, with the help of his brother Mark, started O'ahu Nature Tours. The company slogan is "Conservation through education."

About the Illustrator

Julian P. Hume was born in Ashford, Kent, UK and grew up in Portsmouth, England. From the age of ten he became really interested in natural history, especially extinct species, and at age twelve made a first attempt at putting a book together on the subject, which included writing the text and supplying the illustrations. He continued illustrating on a part-time basis, pausing only to have three daughters, until 1988 when he fortuitously contacted Storrs L. Olson and Helen F. James, two palaeontologists from the Smithsonian Institution who had recently made incredible fossil bird discoveries on the Hawaiian Islands. On the strength of his artwork, Julian was invited on a 1990 field trip to the Hawaiian Islands accompanied by Helen and palynologist, Dave Burney. The trip became a life changing event. Julian became so interested in the science that in 1996 he undertook a degree in palaeontology at the University of Portsmouth, followed by a PhD in the same subject, jointly hosted by the University of Portsmouth and the Natural History Museum, London and Tring. Further trips to Hawai'i took place with Storrs and Helen, but his main area of research was the Mascarene Islands of Mauritius, Réunion, and Rodrigues in the southwestern Indian Ocean, where in particular he has studied the history of the dodo *Raphus cucullatus*. In 2008, he published a critically acclaimed book *Lost Land of the Dodo,* co-authored with Anthony Cheke, on the extinct Mascarene species, in 2012 published *Extinct Birds*, co-authored with Michael Walters, and *Extinct Birds* as sole author in 2017. These three books include Julian's artwork of extinct species. His paintings of extinct birds are produced through a combined knowledge of art and science with each painting executed in the most scientifically accurate way. Julian has contributed both scientifically and artistically to a large number of palaeontological papers, popular books and magazines, and continues working on fossil localities around the world including remote oceanic islands. He has made many new fossil discoveries which have resulted in his description of a number of extinct bird species new to science. (For more information visit www.julianhume.co.uk)

Index

D

E

F

G

H

I

J

K

L

M

N

O

P

R

S

T

U

V

W

X

Z